Emotional :how feelings shape our thinking /
9781524747596

EMOTIONAL

EMOTIONAL

How Feelings Shape Our Thinking

Leonard Mlodinow

PANTHEON BOOKS, NEW YORK

Library of Congress Cataloging-in-Publication Data
Name: Mlodinow, Leonard, [date] author.
Title: Emotional : how feelings shape our thinking /
Leonard Mlodinow.
Description: First edition. New York : Pantheon Books, 2022.
Includes index.
Identifiers: LCCN 2021005986 (print). LCCN 2021005987 (ebook).
ISBN 9781524747596 (hardcover). ISBN 9781524747602 (ebook).
ISBN 9780593316962 (open-market edition).
Subjects: LCSH: Emotions. Reason.
Classification: LCC BF511.M59 2022 (print) | LCC BF511 (ebook) |
DDC 152.4—dc23
LC record available at lccn.loc.gov/2021005986
LC ebook record available at lccn.loc.gov/2021005987

www.pantheonbooks.com

Jacket design by Janet Hansen

Printed in the United States of America
First Edition
2 4 6 8 9 7 5 3 1

In memory of Irene Mlodinow
(1922–2020)

Contents

Contents

Introduction

In some families when a child's misbehavior passes a certain threshold, they give the child a time-out. Or they sit down and talk about why it is important to obey, or not to act out. In other families a parent might give a child a paddling on the rear end. My mother, a Holocaust survivor, wouldn't do any of those things. When I made a big mess or tried to flush the transistor radio down the toilet, my mother would work herself into a frenzy, erupt in tears, and start to scream at me. "I can't take it!" she'd shout. "I wish I were dead! Why did I survive? Why didn't Hitler kill me?"

Her rants made me feel bad. But the strange thing is, as a child, I thought my mother's reaction was normal. You learn many things growing up, but one of the strongest lessons—one that sometimes takes years of therapy to unlearn—is that whatever your parents say about you is correct and whatever happens in your household is the norm. And so I accepted my mother's rants. Sure, I knew that my friends' parents, who hadn't gone through the Holocaust, would not make the Hitler reference. But I imagined them spewing in some analogous manner. "Why did I survive? Why didn't that bus run me down?" "Why didn't that tornado carry me away?" "Why didn't I have a heart attack and drop dead?"

The idea that my mother was an outlier finally occurred to me at dinner one evening when I was in high school. She spoke of a psychiatric appointment she had gone to earlier that day. The visit had been required as part of her application for Holocaust reparations from the German government. The Nazis had confiscated her family's considerable wealth when the war began and left her a pauper. But the payments were apparently based not just on financial considerations. They were based on evidence of emotional problems stemming from what she had endured. My mother had rolled her eyes at having to go to the appointment and was certain that due to her fine mental health she would be denied. But as my brother and I picked at the tasteless boiled chicken on our plates, she told us—indignantly—that the doctor concluded that she indeed had emotional issues.

"Can you believe that?" my mother asked. "He thinks I'm crazy! Obviously, he's the crazy one, not me." And then she raised her voice at me. "Finish your chicken!" she said. I resisted. It has no taste, I complained. "Eat it!" she said. "Someday you might wake up and find that your whole family was killed! And you, with nothing to eat, will have to crawl on your belly through the mud in order to drink stinking, filthy water from mud puddles! *Then* you'll stop wasting food but it'll be too late."

Other kids' mothers lectured them about not wasting food because there were people starving in impoverished faraway lands. My mother told me that *I* might soon be the one desperate to eat. It wasn't the first time my mother had expressed such a sentiment, but this time, backed by my mental image of her wise psychiatrist, I began to question her sanity.

What I know now is that my mother was warning me about the future because she was tortured by her past, terrified that it would repeat. She was telling me that life might look good now, but that was just smoke and mirrors, and would be replaced by a nightmare sometime soon. Not recognizing that her expectation of future cataclysms was rooted in fear, not reality, she believed that her dire expectations were well-founded. As a result, anxiety and fear were never far from the surface.

My father, a former resistance fighter and Buchenwald death camp survivor, had gone through comparable trauma. He and my mother met as refugees soon after the war had ended and for the rest of their lives experienced most life events together. And yet they responded differently, he always being full of optimism and self-confidence. Why did my parents react to events in such varied ways? More generally, what *are* emotions? Why do we have them, and how do they arise in our brains? How do they affect our thoughts, judgments, motivation, and decisions, and how can we control them? These are the questions I will address in this book.

The human brain is often compared to a computer, but the information processing that this computer executes is inextricably intertwined with the deeply mysterious phenomenon we call feelings. We've all felt anxiety, fear, and anger. We've felt rage, despair, embarrassment, loneliness. We've felt joy, pride, excitement, contentment, lust, and love. When I was a child, scientists had little idea of how those emotions are formed, how one can manage them, what purpose they serve, or why two people—or the same individual at different times—may respond to the same triggers in quite disparate ways. Scientists back then believed that rational thought was the dominant influence on our behavior and that when emotions played a role they were likely to be counterproductive. Today we know better. We know that emotion is as important as reason in guiding our thoughts and decisions, though it operates in a different manner. While rational thought allows us to draw logical conclusions based on our goals and relevant data, emotion operates at a more abstract level—it affects the importance we assign to the goals and the weight we give to the data. It forms a framework for our assessments that is not only constructive but necessary. Rooted in both our knowledge and our past experience, emotion changes the way we think about our present circumstances and future prospects, often in subtle but consequential ways. Much of our understanding of how that works has come from advances in just the last decade or so, during which there has been an unparalleled explosion of research in the field. This book is about that revolution in our understanding of human feelings.

Before the current burst of research into emotion, most scientists understood our feelings within a framework that goes all the way back to the ideas of Charles Darwin. That traditional theory of emotion embraced a number of principles that seem intuitively plausible: that there is a small set of basic emotions—fear, anger, sadness, disgust, happiness, and surprise—that are universal among all cultures and have no functional overlap; that each emotion is triggered by specific stimuli in the external world; that each emotion causes fixed and specific behaviors; and that each emotion occurs in specific dedicated structures in the brain. This theory also encompassed a dichotomous view of the mind that goes back at least to the ancient Greeks: that the mind consists of two competing forces, one "cold," logical, and rational and the other "hot," passionate, and impulsive.

For millennia these ideas informed thinking in fields from theology to philosophy to the science of the mind. Freud incorporated the traditional theory into his work. John Mayer and Peter Salovey's theory of "emotional intelligence," popularized by the 1995 book of that name by Daniel Goleman, is in part based on it. And it is the framework for most of what we think about our feelings. But it is wrong.

Just as Newton's laws of motion were superseded by quantum theory when science developed the tools that revealed the atomic world, so too is the old theory of emotion now giving way to a new view, thanks in large part to extraordinary advances in neuroimaging and other technologies that have allowed scientists to look into, and experiment upon, the brain.

One set of techniques developed in the past few years allows scientists to trace the connections among neurons, creating a kind of circuit diagram for the brain called the "connectome." The connectome map allows scientists to navigate the brain in a way that was never before possible. They can compare essential circuits, fly into specific regions of the brain to explore the cells that they comprise, and decipher the electrical signals that generate your thoughts, feelings, and behaviors. Another advance, optogenetics, lets scientists take *control* of individual neurons in an animal's brain. By selectively stimulating them, scientists have been able to uncover the micropatterns of brain activity that produce certain mental states, such as fear, anxiety, and depression. A third technology, transcranial stimulation, employs electromagnetic fields or currents to stimulate or inhibit neural activity in precise locations in the human brain with no permanent effects on the experimental subject, helping scientists to assess the function of those structures. These and other techniques and technologies have imparted so much insight, and given rise to so much new work, that a whole new field of psychology has emerged, called "affective neuroscience."

Founded on the application of modern tools to the age-old study of human feeling, affective neuroscience has reshaped the way scientists view emotion. They've found that while the old viewpoint offered what seemed like plausible answers to basic questions about feelings, it didn't accurately represent the way the human brain operates. For example, each "basic" emotion is not really a single emotion but actually a catchall term for a spectrum or category of feelings, and those categories are not necessarily distinct from one another. Fear, for instance, comes in different flavors and can in some instances be difficult to distinguish from anxiety.[1] What's more, the amygdala, long thought of as our "fear" center, actually plays a key role in several emotions and, conversely, is not necessary for all types of fear. Scientists today have also expanded their focus far beyond the "basic" five or six emotions to include dozens of others, such as embarrassment, pride, and other so-called social emotions, and even feelings that used to be thought of as drives, such as hunger and sexual desire.

In the domain of emotional health, affective neuroscience has taught us that depression is not a single disorder but rather a syndrome comprising four different subtypes, susceptible to different treatments and having different neural signatures. Researchers have used the new insights to develop a phone app that can help alleviate depression in a quarter of depressed patients.[2] In fact, scientists can now sometimes determine in advance, through a brain scan, whether a depressed person will benefit from psychotherapy rather than requiring drugs. Potential new treatments for emotion-related conditions from obesity to smoking addiction to anorexia are also being studied.

Fueled by such triumphs, affective neuroscience has come to be one of the hottest fields in academic research. It has become prominent in the National Institute of Mental Health's research agenda and in many institutions that are not commonly thought of as mind focused, such as the National Cancer Institute.[3] Even institutions that have little to do with psychology and medicine, such as computer science centers, marketing organizations, business schools, and the Kennedy School of Government at Harvard, are now devoting resources and jobs to this new science.

Affective neuroscience has important implications for the place of feelings in our everyday lives and in the human experience. Said one leading scientist, "Our traditional 'knowledge' about emotion is being questioned at the most fundamental level."[4] Said another leader in the field, "If you are like most people, you feel convinced that because you have emotions, you know a lot about what emotions are, and how they work . . . you are almost certainly wrong."[5] According to a third, we are "in the midst of a revolution in our understanding of emotion, the mind, and the brain—a revolution that may compel us to rethink such central tenets of our society as our treatments for mental and physical illness, our understanding of personal relationships, our approaches to raising children, and ultimately our view of ourselves."[6]

Most important, where we once believed that emotion was detrimental to effective thought and decisions, we now know that we can't make decisions, or even think, without being influenced by our emotions. And though—in our modern societies that are so different from the surroundings in which we evolved—our emotions are sometimes counterproductive, it is far more often the case that they lead us in the right direction. In fact, we'll see that without them we'd have difficulty moving in any direction at all.

WHAT'S AHEAD

Given their experiences in the Holocaust, my parents might not seem typical. But in a fundamental way we are all just like them. Deep within our brains, as in theirs, our shadowy unconscious mind is continuously applying the lessons of our past experience to predict the consequences of our current circumstances. In fact, one way to characterize a brain is as a prediction machine.

Hominids evolving on the African savanna faced constant decisions regarding food, water, and shelter. Is that rustling up ahead caused by an animal I can eat or one that wants to eat me? Animals that were better at analyzing their surroundings were more likely to survive and reproduce. Toward that end, given any circumstance, the job of their brains was to use their sensory input and past experience to decide on a set of possible actions and then, for each possible action, to forecast the likely outcomes. Which action is least probable to lead to death or injury, and most likely to provide nutrition, water, or some other contribution to their survival? In the pages that follow, we will look at how emotion influences those calculations. We'll look at how emotion arises, the role of our feelings in creating our thoughts and decisions, and how we can harness our feelings to thrive and be successful in the modern world.

In part I, I will describe our current knowledge of how emotions evolved and why. An understanding of the role of emotion in our basic blueprint for survival will reveal a lot about how we respond to situations; why we react with anxiety or anger, love or hate, happiness or sadness; and why we sometimes act inappropriately or lose control of our emotions.

We'll also explore the concept of "core affect," the mind-body state that subliminally informs all your emotional experience, influencing not just the emotions you will feel in any given situation but also your decisions and reactions to events—one reason that on different occasions the same circumstance can create in you quite different emotional responses.

Part II will look into the central role of emotion in human pleasure, motivation, inspiration, and determination. Why, given two tasks of comparable interest, difficulty, and importance, might one of them seem so hard to achieve, the other easy? What factors affect the intensity of your desire to accomplish something? Why, in similar situations, do you sometimes press on with herculean effort and at other times immediately give up? And why are some individuals more prone to pressing on, others to dropping out?

Part III will explore emotional profile and emotion regulation. We each have tendencies to react with certain emotions and a disinclination to react with others. Scientists have developed questionnaires you can take to assess your own tendencies in several of those major dimensions, and I will present those in chapter 8. Chapter 9 will examine the burgeoning field called "emotion regulation," time-tested strategies for emotion management that have recently been studied and corroborated through rigorous scientific research. Once you've understood where your feelings come from, how can you take charge of them? What makes that more difficult for some people than for others?

We all spend time deliberating about what restaurant to go to or what film to see, but we don't necessarily devote time to pondering ourselves, to examining what we feel and why. Many of us were actually raised to do the opposite: we were taught to suppress emotion; we were taught not to feel. But though we can suppress emotion, we can't "not feel." Feelings are part of being human and of interacting with other humans. If we're not in touch with them, we are not in touch with ourselves, and that will hamper us in our dealings with others and doom us to make judgments and decisions without a full understanding of the origin of our thinking.

As I write this, my mother is ninety-seven years old. She has mellowed, but at her core she never changed. Having studied the new theory of emotion, I've gained insight into her behavior. More important, I've gained insight into mine, for to know yourself is a first step toward both acceptance and change, if you desire it. My hope is that this journey through the science of emotion will debunk the myth that emotions are counterproductive and offer a new understanding of the human mind that can help you navigate your world of feelings and gain control and power over them.

PART I

What Is Emotion?

1

Thought Versus Feeling

On the morning of Halloween 2014, a strange aircraft ascended high into the skies above the barren Mojave Desert. The custom-built carbon-fiber plane was essentially twin cargo jets flying side by side, joined at the wing. Suspended from that monstrous carrier vessel was a smaller plane dubbed the *Enterprise*—an homage to *Star Trek*. The aim was for the cargo jet to carry the *Enterprise* to an altitude of fifty thousand feet, from where it would be dropped, briefly fire its engines, and then glide to a landing.

The planes belonged to Virgin Galactic, the company created by Richard Branson to carry "space tourists" into suborbital flight. By 2014, more than seven hundred spaceship tickets had been sold, at $200,000 to $250,000 each. This was the thirty-fifth such test flight but only the fourth in which the *Enterprise* was meant to fire up its rocket, which had just been redesigned to make it more powerful.

The ascent went well. The pilot David Mackay launched the *Enterprise* from the underside of his carrier plane at the appointed moment. Then his eyes panned across the sky, searching for the plume of the *Enterprise*'s rocket engine. He couldn't spot it. "I remember looking down and thinking, 'Well that's strange,'" recalled Mackay, experienced enough to be wary of anything unexpected.[1] But all was well. Out of his line of sight, the spaceship had indeed fired its rocket and in about ten seconds accelerated through the sound barrier. The mission was unfolding without incident.

The *Enterprise* was captained by a test pilot named Peter Siebold, with almost thirty years of flying experience. His co-pilot, Michael Alsbury, had previously worked with eight different experimental aircraft. In some ways, the two men were quite different: while Siebold could strike co-workers as aloof, Alsbury was always friendly and known for his sense of humor. But strapped into their seats atop the rocket, they functioned as a unit, each of their lives dependent on the actions of the other.

Just before reaching the speed of sound, Alsbury unlocked the ship's air-braking device. The brake was crucial for controlling the spaceship's orientation and speed while dropping back to earth, but it wouldn't be needed for another fourteen seconds, and Alsbury had unlocked it before he should have. The National Transportation Safety Board would later criticize the Scaled Composites unit of Northrop Grumman, which designed the vehicle for Virgin, because it did not guard against such human slipups by providing a fail-safe system to prevent premature unlocking.

Unlike Virgin Galactic, government-sponsored space initiatives call for "two-failure tolerance." That means putting in place safeguards to protect against two separate and unrelated simultaneous problems—two human errors, two mechanical errors, or one of each. The Virgin team was confident that its extraordinarily well-trained test pilots wouldn't make such mistakes, and eliminating safeguards had certain advantages. "We don't have all the constraints a government organization like NASA would," one team member told me. "So we can get things done a lot faster."[2] But on that Halloween morning, the lock disengagement was no harmless mistake.

With the lock off prematurely, the force of the atmosphere caused the brake to deploy early, even though Alsbury never threw the second switch to deploy it. As the brake swung into position, the still-firing rocket placed tremendous stress on the plane's fuselage. Four seconds later, traveling at 920 miles per hour, the ship ripped apart. From the ground, it looked like a massive explosion.

Siebold, still attached to his ejection seat, was thrown from the plane. Traveling faster than sound, he was in an atmosphere where the temperature of the air around him was minus seventy degrees Fahrenheit, and there was just one-tenth the oxygen present at sea level. Still, he somehow managed to unbuckle himself, after which his parachute automatically opened. Upon rescue, he had no memory of the experience. Alsbury wasn't so lucky. He died instantly when the plane broke apart.

EMOTIONS AND THOUGHT

The long string of well-rehearsed procedures called for when a pilot tests a new plane are normally executed so smoothly that it's easy to think of them as rote and mechanical. But that view is profoundly misguided. When the *Enterprise* was dropped from its mother ship and started to fire its ferocious rocket engine—as planned—the physical circumstance of its pilots was suddenly disrupted. It's hard to imagine what that felt like, but a rocket is really a controlled exploding bomb, and a controlled explosion is still an explosion. It's a terribly violent event, and the *Enterprise* was relatively flimsy—a mere twenty thousand pounds, loaded, as compared with the space shuttle's four million. And so the ride is much different. If flying in the space shuttle is like racing down the highway in a Cadillac, piloting the *Enterprise* was like driving 150 miles per hour in a go-kart. The souped-up rocket's firing subjected the *Enterprise* pilots to a colossal roar, savage shaking and vibration, and fierce stresses of acceleration.

Why did Alsbury throw the switch when he did? The flight was proceeding as planned, so it's not likely he was panicking. We can't know what his reasoning was, nor perhaps did he. But in the anxious state that comes from a highly stressful physical environment, we process data in a manner that is hard to predict from practice runs in flight simulators. This was more or less the conclusion of the National Transportation Safety Board about the events on the *Enterprise*. Speculating that Alsbury, lacking recent flight experience, might have been unusually stressed, the NTSB posited that he committed the misjudgment due to the anxiety caused by time pressure and the ship's strong vibration and forces of acceleration, which he hadn't experienced since his last test flight eighteen months earlier.

The story of the *Enterprise* illustrates how anxiety can lead to a bad decision, as it surely sometimes does. In our ancestral environment, there were many more life-threatening dangers than we typically face in civilized life, and so our fear and anxiety reactions, in particular, may at times seem overblown. Such cases, as exemplified by the *Enterprise* saga, are what, over the centuries, gave emotion a bad name.

But stories of emotions causing problems are often sensational, as this one was, while tales of emotions operating as they should tend to be mundane. It is the malfunctions that stand out in the telling, while a properly functioning system can easily go unheralded. There were, for example, thirty-four successful prior test flights of the *Enterprise*. In each of those, both the plane and its pilots operated as planned, controlled by a miraculous marriage of modern technology and the smooth interplay of the rational and emotional human brain, and none of them made the news.

A case that hit closer to home for me concerned a friend who lost his job, and therefore his health insurance. Knowing the cost of decent medical care, he became anxious about his health. What if he got sick? He could go broke. That anxiety affected his thinking—if he had a sore throat, he didn't ignore it or dismiss it as sniffles as he'd have previously done. Instead, he'd fear the worst: Was it throat cancer? As it turned out, his anxiety over his health saved his life. For one of the things he had never paid attention to, but now began to worry about, was a mole on his back. For the first time in his life, he went to a dermatologist and had it checked out. It was an early-stage cancer. He had it removed, and it never recurred—a man rescued by anxiety.

The moral of this pair of stories is not that emotions help or impede effective thinking but rather that *emotions affect thinking*: our emotional state influences our mental calculations as much

as the objective data or circumstances we are pondering. As we'll
see, that is usually for the best. It is the exception and not the rule
when the effect of emotion proves counterproductive. In fact, as
we explore the purpose of emotion in this and the next several
chapters, we'll see that, indeed, if we were "free" from all emotion,
we would hardly be able to function because our brains would
have to be hopelessly cluttered with rules governing the simple
decisions we must constantly make to react to the everyday circumstances of life. But for now, let's focus, not on the detriments
or benefits of emotion, but on emotion's role in the way our brains
analyze information.

Emotion states play a fundamental role in the biological information processing of all creatures, in mammals as well as simple
insects, and in the actions they take as a result. In fact, the very
process that went awry in the *Enterprise* disaster was mirrored
in a controlled experiment in which honeybees were put into an
extreme situation eerily parallel to that of the Virgin pilots.[3] The
researchers in that study were interested in how such simple creatures might respond to being in a chaotic and dangerous situation,
and so they subjected them to sixty seconds of high-speed shaking.

How do you subject bees to "high-speed shaking"? After all,
if you simply capture bees in a vessel and shake it, they can hover
inside so what you'll have is bees flying around a shaking jar, not
bees that themselves are shaken. To circumvent such issues, these
researchers immobilized the bees by strapping them into tiny bee
harnesses, adding to the similarity of their plight to that of the
Virgin pilots, who were also securely strapped down and immobilized as their vessel violently quaked. In the case of the bees, the
harnesses were made from a small length of a plastic straw or
other tubing, cut in half along its length. Each bee was cooled
down to become briefly inactive as it was laid inside the half tube
and secured with duct tape.

After the shaking, the scientists tested the bees' decision making. They presented them with a task that required them to dis-
criminate between various odors they had previously been exposed
to. In those prior exposures, the bees had learned which odors

signaled a pleasant treat (a sucrose solution) and which denoted an unpleasant liquid (quinine). Now, after being shaken, each bee was again presented with fluids to sample. The bees were put in a position in which they could choose, based on the odor associations they had learned earlier, whether to drink each sample or pass it up.

But these post-shaking samples were not purely pleasant or unpleasant; they were two-to-one mixtures, either predominantly pleasant sucrose or mostly the unpleasant quinine. The two-to-one sucrose-quinine mixture was still pleasant to the bees, and the two-to-one quinine-sucrose still unpleasant, but now the odors were ambiguous. A bee, when presented with each mixed sample, had to decide whether the ambiguous smell signaled a pleasant treat or an unpleasant surprise. The scientists were interested in the question, would the prior shaking affect the bees' assessments of the odors, and if so, how?

Anxiety in bees, as in humans, is a reaction to what affective neuroscientists call a "punishing" environment. In the case of the *Enterprise* and the bees, that needs no explanation, but more generally it means a circumstance in which a threat to comfort or survival might reasonably be expected.

Thinking in an anxious state, scientists have found, leads to a pessimistic cognitive bias; when an anxious brain processes ambiguous information, it tends to choose the more pessimistic from among the likely interpretations. Your brain becomes overactive in perceiving threats and tends to predict dire outcomes when faced with uncertainty. It's easy to understand why brains might be designed that way; being in a punishing environment, one would be wise to interpret ambiguous data as being more threatening, or less desirable, than one might if the surroundings were safe and pleasant.

That pessimistic bias in judgment is just what the scientists found. The shaken bees passed up the two-to-one sucrose-quinine solution significantly more often than a control group that had not been shaken: the shaking had influenced the bees to interpret the ambiguous smell as signaling an undesirable liquid. One might be tempted to describe the scientists' result by saying that the shaken bees made more "mistakes" than the control group. That would fit the "emotions impede good decision making" narrative, but this controlled experiment makes it clear that what was really going on was a reasonable threat-justified shift in the bees' judgment.

The shaking-induced anxiety had to also affect the judgment of the *Enterprise* pilots. People, like bees, get anxious when they experience turbulence in their external world, and it affects their information processing in a similar way. That's even true physiologically; anxious bees have lower levels of the neurotransmitter hormones dopamine and serotonin in their hemolymph (bee blood), just as humans do when under anxiety.

"We show that the bees' response to a negatively valenced event has more in common with that of vertebrates than previously thought," the researchers wrote. "[This] suggests that honeybees could be regarded as exhibiting emotions." Though the scientists were saying that the behavior of bees reminded them of the behavior of people, to me the conditions of the pilots—being vibrated and shaken—reminded me of the bees. On some deep level of existence, we and bees have a surprising and revealing commonality with regard to the way we process information: it is not just a "rational" exercise; it is deeply intertwined with emotion.

Affective neuroscience tells us that biological information processing cannot be divorced from emotion, nor should it be. In humans, that means that emotion is not at war with rational thought but rather a tool of it. As we'll see in the chapters that follow, in the thinking and decision making in endeavors ranging from boxing to physics to Wall Street, emotions are a crucial element of success.

OUTGROWING PLATO

Because our mental processes are so mysterious, their nature has occupied thinkers since long before we even understood that the brain is an organ. One of the first and most influential to contemplate them was Plato. He envisioned the soul as a chariot pulled by two winged horses guided by a charioteer. One horse was a "crooked lumbering animal . . . of a dark color, with grey eyes and blood-red complexion . . . hardly yielding to whip and spur." The other was "upright and cleanly made . . . a lover of honor . . . and the follower of true glory; he needs no touch of the whip, but is guided by word and admonition only."

Much of what we talk about when we speak of how emotion motivates behavior is illustrated by Plato's chariot. The dark horse stands for our primitive appetites—for food, drink, sex. The other horse symbolizes our higher nature, our emotional drive to achieve goals and accomplish great things. The charioteer represents the rational mind, trying to harness both horses for its own purposes.

In Plato's view, a competent charioteer would work with the white horse to restrain the dark horse and train them both to keep moving upward. Plato believed that the deft charioteer also listens to the desires of both horses and works to channel their energy and achieve harmony between them. In Plato's thinking the task of the rational mind is to take stock of and control our drives and desires and, in light of our goals, to choose the best course. Although we now know that it's misguided, that division between the rational and the nonrational mind became one of the main themes of Western civilization.

Though Plato saw our emotions and rationality working harmoniously, in the centuries after Plato those two aspects of our mental life came to be seen as working in opposition to each other. Reason was viewed as superior and even holy. Emotions were to be avoided or contained. Later Christian philosophers accepted that view in part. They grouped human appetites, lusts, and passions as sins a virtuous soul should seek to avoid, but identified love and compassion as virtues.

The term "emotion" grew out of the work of Thomas Willis, a seventeenth-century London doctor. He was also an enthusiastic anatomist, and if you died under his care, there was a fair chance he would dissect you. In situations of life or death, it couldn't have been comforting to know that your doctor wins either way. But Willis also had another source of cadavers: he had obtained permission from King Charles I to perform autopsies on hanged criminals.[4]

In the course of his research, Willis identified and named many of the brain structures that we still study today. More important, he found that the deviant behaviors of many criminals could be traced to specific features of those structures. Later physiologists built upon Willis's work to examine reflexive responses in animals. They found that expressions such as the recoil of fright come from purely mechanical processes governed by nerves and muscles; they involved some sort of motion. Soon the word "emotion," derived from the Latin *movere*, "to move," appeared in both English and French forms.

It took a couple centuries to get the "motion" out of "emotion." The modern use of the term first appeared in a set of lectures published in 1820 by an Edinburgh professor of moral philosophy named Thomas Brown. The book of lectures was wildly popular and went through twenty editions over the following several decades.[5] Thanks to John Gibson Lockhart, who was Sir Walter Scott's son-in-law, we have some idea of what the scene was like when Brown delivered these lectures: Lockhart included an account of one in a fictionalized portrait he wrote describing Edinburgh society. In it, Brown arrives "with a pleasant smile upon his face, arrayed in a black Geneva cloak over a snuff-coloured coat and a buff waistcoat," his manner of speaking "distinct and elegant," his ideas enlivened by quotations from the poets.

In his lectures Brown proposed a systematic study of emotion. While an excellent idea, it faced immense hurdles. This was an era in which Auguste Comte, sometimes called the first philosopher of science, examined each of the six "fundamental" sciences—mathematics, astronomy, physics, chemistry, biology, and sociology—but did not include psychology. And for good reason: while John Dalton was discovering the basic laws of chemistry, and Michael Faraday the principles of electricity and magnetism, there wasn't yet a fundamental science of the mind. Brown had wanted to change that. Redefining emotion as "all that is understood by feelings, states of feelings, pleasures, passions, sentiments, affections," he grouped emotions into categories and proposed that they be studied scientifically.

Brown had many great traits as a philosopher-scientist, but staying alive was not one of them. He collapsed while lecturing in December 1819. His doctor examined him and then sent him to London for a "change of air." He died there on April 2, 1820, shortly before his book came out. He was forty-two. Though Brown never knew the impact of his ideas, his lectures guided scholars' thinking about emotion for years to come. Today he is a little-known figure, his grave in a state of disrepair. But for decades after his death he was well celebrated for his insights into the human mind.

The next great leap in the study of emotion came from Charles Darwin, who started to ponder the topic upon returning from his voyage on the *Beagle* in 1836. Darwin hadn't always been interested in emotion, but as he began to create his theory of evolution, he scrutinized all aspects of life to try to understand how they fit into his puzzle. Emotion was one he had trouble with. If, as was then generally accepted, emotions were counterproductive, why would they have evolved? Today we know they are not counterproductive, but to Darwin the quandary was a test for natural selection. How do these apparently disadvantageous emotions fit into animal behavior? Despite the dearth of prior work in the field, Darwin was determined to find the answer. It took him decades to formulate his explanation.

EMOTIONS AND EVOLUTION

Some of the most detailed studies Darwin performed were on nonhuman animals because the function of an emotion is often clearer in simpler organisms. Anxiety, for example, plays a complex and shifting role in our lives that is far different from its role in the natural world from which we evolved, but its constructive role in the animal world is more straightforward and easy to read. Take the ruddy duck, for example.

Because evolution depends upon successful mating, the genitals of every species are adapted to meet the particular circumstances it faces. In the case of the ruddy duck, the female genitals evolved to inhibit access to undesirable males, preventing impregnation unless the female assumes a posture that enables the male to penetrate fully. That allows the female to be selective about whom she mates with. The males, of course, evolved in response.

During the summer male ruddy ducks have dull plumage, similar to that of the female, which makes them inconspicuous to predators. But as the winter mating season approaches, they temporarily don the ruddy duck equivalent of a Rolex watch and gold necklace—their feathers take on a rich chestnut hue and their bill turns a bright blue—advertising themselves to the picky females. In addition to flaunting their duck bling, they perform unusual courtship displays in which they stick their tails straight up while striking their beaks against their inflated necks. The ducks' bright plumage and beaks are riskier than their off-season camouflage, but maybe that's the point: it sends females a message of physical prowess, that the individual is fit enough it needn't fear being noticed.

The system works pretty well, but there is one more necessary adjustment. Because the female genitals are difficult to access, to successfully mate the male genital needs to be extra-long—sometimes as long as the creature's body itself. Because a genital like that is difficult to lug around, like the bright feathers it is shed when the time to mate is over and then regrown each year.

So far as we know, the ruddy ducks have no anxiety about this annual shedding of their penises, but what they do have anxiety about is the threat of violence. Male ruddy ducks can be bullies, with larger ducks bullying the smaller ones. The frequency of physical conflict is diminished, however, because anxiety about being attacked spurs the weaker ones to shed their colored plumage more quickly and to grow a far tinier sex organ. That has the effect of making them less of a competitive threat at mating time and therefore less of a target for aggression. This social dynamic plays an evolutionary role similar to the establishment of dominance hierarchies in primates and other social animals: it allows for the resolution of conflict without costly fighting that can result in serious injury or death and serves to maintain order among pack members.

No one knows to what extent the ducks consciously "feel" the emotion of anxiety, but scientists can measure the biochemical changes within their bodies that result from it. The research journal *Nature* summed this all up in a headline, "Sexual Competition Among Ducks Wreaks Havoc on Penis Size."[6] By effectively ceding the choice of mate to the more powerful and minimizing the potential for wasteful violence, that "havoc" confers an evolutionary benefit on the species. At least in this instance, anxiety's positive role in the dance of evolution is clear.

The evolutionary roles of many human emotions are also fairly clear. Consider the feelings we have about that by-product of mating that we call babies. About two million years ago our ancestor *Homo erectus* evolved a much larger skull, which allowed for expansion of the frontal, temporal, and parietal lobes of the brain. That gave us, like a new smartphone model, a great boost in our computing power. But it also caused problems, because, unlike a smartphone, a brand-new human has to slide down the birth canal of an older human, and it has to be supported by the mother's own metabolic activity until that blissful moment. As a result of such challenges, human babies make their exit earlier than is normal for primates—a human pregnancy would have to last eighteen months to allow the brain of a human child to be as developed as that of a chimpanzee when it is born, by which time the baby would be too large to exit the birth canal. The earlier exit solves some problems, but it causes others. Because the human brain at birth is not very well developed (only 25 percent of adult size, as opposed to 40–50 percent for an infant chimp), human parents are burdened with a child who will remain helpless for many years—about twice as many years as a baby chimp.[7]

Caring for that helpless child is a major life challenge. Not long ago, I had lunch with a friend who had, at the birth of his child fifteen months earlier, become a stay-at-home dad. My friend had played college football and later been the CEO of a start-up company. Neither of those challenges wore him down. But at our recent lunch he was sullen, tired, hunched over with a sore back, and limping. In other words, on him, stay-at-home fatherhood had had the same effect as a mild case of polio.

My friend is not atypical. Human children require an enormous amount of care. The job of providing that care is one of the least appreciated professions in Western society, but it takes its toll. Before their first child is born, some people think having a child will be one big party. What they don't realize is that with that big party comes a hangover—their duties as the child's cleanup crew, caterer, and security guard.

Why do we get up three times a night to feed our kids? Why do we take pains to wipe up their poop, and remind ourselves to lock the cabinet that contains the silver polish that looks like a bottle of Gatorade? Evolution has provided a motivating emotion for all that work: parental love.

Each of our emotions, when it occurs, alters our thinking in a manner that fulfills some evolutionary purpose. Our parental love is as surely a cog in the machine of human life as is mating anxiety in the life of the ruddy duck. That we love our children because evolution has manipulated us to do so doesn't diminish that love. It merely reveals the origin of that gift that so enriches our lives.

Darwin, trying to puzzle out the role of emotion, didn't have access to the background knowledge and technologies we do today, and he never studied the ruddy duck (they are native to North America). But he did study in great detail the plumage, skeleton, bills, legs, wings, and behavior of numerous other wild ducks. He also interviewed pigeon and livestock breeders. And he examined an ape, an orangutan, and monkeys in the London Zoo.

Believing that he could glean insight into the purpose of emotions by focusing on the outward signs—those muscle movements and configurations, especially in the face, that had inspired the coining of the term itself—he took copious notes on animals' seeming humanlike expressions of feeling. He became convinced that animals "are excited by the same emotions as ourselves" and that the outward signs of emotion served to communicate those feelings, enabling a kind of mind reading among animals that lacked the capacity for language.[8] Dogs might not cry at the end of *Romeo and Juliet*, but in his dog's gaze Darwin believed he saw the emotion of love.

Darwin studied emotions in humans, too, again concentrating on their physical manifestation. He circulated a questionnaire among missionaries and explorers across the world, asking about emotional expression in different ethnic groups. He examined hundreds of emotional photographs of actors and babies. He documented the smiles and frowns of his own infant son, William. His observations led him to believe that each emotion yields a characteristic and consistent expression across all human cultures—just as he'd observed across varying species of other mammals. Smiles, frowns, eye widening, our hair standing on end, Darwin believed, all derive from physical displays that proved useful in the earlier stages of our species' evolution. For example, when it faces an aggressive rival, a baboon snarls to signal its readiness to fight. A wolf can also snarl or send the opposite message by rolling submissively on its back to telegraph willingness to back down.

Darwin concluded that our various emotions were passed down to us from ancient animal ancestors in whose lives each emotion played a specific and necessary role. That was a revolutionary idea, a profound departure from the pervasive millennia-old view that emotions are fundamentally counterproductive.

Yet Darwin also believed that sometime in the course of our evolution we humans developed a superior method of information processing—our rational mind, a "noble" and "god-like intellect" that could override our irrational emotions—and so he wrongly believed that emotions had ceased to have a constructive function.[9] Our emotions, in Darwin's view, were mere remnants of a previous stage of development, like our tailbone or appendix—useless, counterproductive, at times even dangerous.

THE TRADITIONAL VIEW OF EMOTION

Darwin finally published his conclusions in his 1872 book, *The Expression of the Emotions in Man and Animals*. It became the most influential work on emotion since Plato and, in the century that followed, inspired the theory of emotion—the "traditional" theory—that until recently dominated our ideas about the subject. The fundamental tenets of the traditional theory were that there are a handful of basic emotions shared by all humans; that those emotions have fixed triggers and cause specific behaviors; and that each arises in some dedicated structure within the brain.

With its roots in Darwinian thinking, the traditional theory of emotion is closely tied to a view of the brain, and its evolution, that is called the "triune model." Carl Sagan popularized the model in his best-selling book *The Dragons of Eden*, and Daniel Goleman relied on it in his 1995 best seller, *Emotional Intelligence*. As presented in most textbooks published between the 1960s and about 2010—and still in many today—the triune model states that the human brain is made of three successively more sophisticated (and evolutionarily newer) layers. The deepest is the reptilian or lizard brain, the seat of your basic survival instincts; the middle layer is the limbic, or "emotional," brain that we inherited from prehistoric mammals; and the outermost and most sophisticated layer is the neocortex, said to be the source of our power of rational thought. These are essentially Plato's dark horse, white horse, and charioteer.

The reptilian brain, according to the triune model, encompasses our brain's most ancient structures, inherited from reptiles, the most instinctive of vertebrates. These structures control our body's regulatory functions. For example, when your blood sugar is low, they produce hunger.

If you are hungry and spot prey, though a reptile will strike, a mammal such as a cat might instead toy with it. A human might pause at the sight of the food source and relish the moment. According to the triune model, the source of these more complex behaviors is the limbic brain, absent in the reptile. The limbic brain is said to be the seat of the basic emotions described by the traditional theory—fear, anger, sadness, disgust, happiness, and surprise.

Finally, the neocortex, which sits atop the limbic structures, is the source of our reason, our abstract thinking, our language and planning abilities, and our conscious experience. It is divided into two halves, or hemispheres, each in turn divided into four lobes—the frontal, parietal, temporal, and occipital—which have different sets of functions. For example, vision is centered on the occipital lobe, while the frontal lobe contains areas enabling abilities that are enhanced in or unique to our species, such as complex language processing in the prefrontal cortex and social processing in the orbitofrontal cortex (a part of the frontal lobe).

The hierarchy of the triune model goes hand in hand with the traditional theory of emotion. It states that the neocortex, our intellectual center, has little or no role in creating our emotional life. Instead, it serves to regulate any counterproductive impulses that arise from it. Emotion, in this scheme, comes from the lower layers. There, each emotion is sparked by specific stimuli in the external world, almost like a reflex. Once triggered, each emotion produces a characteristic pattern of physical changes. These involve different sensations and bodily reactions such as patterns of heart rate and breathing and the configuration of facial muscles. A particular situation, in that view, would almost always result in a given emotional response, and almost everyone—in all cultures—would have that same response, unless the structures involved in creating the emotion are damaged.

The triune model puts emotion, brain structure, and evolution in a neat package. The only problem with it is that it is not accurate—at best a vast oversimplification. Though neuroscientists still sometimes use it as a shorthand, misunderstandings will result if one takes it literally. For one, it does not account for the great deal of communication that occurs between the layers. If a food smell generates disgust in the limbic brain, for example, it may transmit that to the reptilian brain, leading to an impulse to vomit, and to the neocortex, which may cause you to step back from the object. What's more, the generation of various emotions in the brain does not seem to be focused in one area or another, as was once thought, but is instead much more widely distributed. There is also anatomical overlap between the layers, making the very classification as reptilian, limbic, and neocortical rather problematic. The orbitofrontal cortex, for instance, is often thought of as a limbic structure.[10] And, finally, evolution doesn't work in the manner portrayed by the triune model. Though various structures in the three layers might have originated in different evolutionary eras, the older structures continued to evolve as the newer ones developed—as did their function and, more generally, their role in the brain's organization. "Adding [layer upon layer] is almost certainly not the way the brain has evolved," said Terrence Deacon, a neuroanthropologist at Berkeley.[11]

The traditional view of emotion, though still common in popular culture, is no more valid than the triune model that seemed to support it. It, too, is only a very rough approximation, and is often misleading. Like Newton's laws of motion, the traditional view of emotion matches our superficial and intuitive understanding but fails if you have the tools to look more closely. In the early twentieth century, new technologies gave scientists the ability to observe nature on a deeper level than Newton could and revealed Newton's "classical mechanics" to be a mere facade. Similarly, twenty-first-century technology has provided scientists with the means to look beyond the superficial aspects of emotion, with the result that the traditional theory of emotion has also been proven wrong.

SAVED BY EMOTION

Shortly after midnight on August 30, 1983, Korean Air Lines Flight 007 took off from John F. Kennedy International Airport in New York, bound for Seoul. The flight was carrying 23 crew members and 246 passengers, including the ultraconservative U.S. congressman Larry McDonald of Georgia, on his way to attend ceremonies celebrating the anniversary of the U.S.–South

Korea Mutual Defense Treaty. According to the *New York Post*, the former president Richard Nixon was to have been seated next to McDonald but decided not to go.

After refueling in Anchorage, the plane, a Boeing 747, took off again and headed southwest toward Korea. About ten minutes later, it began to deviate to the north. Half an hour after that, an automated military radar system at King Salmon, Alaska, detected the plane about twelve miles north of where it should have been, but military personnel were not notified. KAL 007 continued on the same heading for the next five and a half hours.

At 3:51 a.m. local time, the plane entered the restricted airspace of the Soviet Kamchatka Peninsula.[*] After tracking the plane for an hour, the Soviet defense forces sent three Su-15 fighters and a MiG-23 to make visual contact. "I saw two rows of windows and knew that this was a Boeing," the lead pilot later said. "But for me this meant nothing. It is easy to turn a civilian type of plane into one for military use."[12] He fired warning missiles toward the plane, expecting its pilot to recognize the military interception and allow them to escort it to a landing. But the missiles flew past the Boeing undetected. Unfortunately, at the same time, the captain of the KAL flight was radioing to the Tokyo-area air traffic control, requesting permission to ascend to a higher flight path to save fuel. Permission was granted. When the Boeing slowed and began to climb, the Soviet pilot interpreted the action as an uncooperative evasive maneuver. He felt uneasy about attacking what might be a civilian aircraft, but he followed military protocol and responded by firing two air-to-air missiles at the plane. The 747 was hit, spiraled down, and crashed into the ocean. No one survived.

NATO responded to the attack with a series of military exercises. These raised cold war tensions between the United States and the Soviet Union, which were already at a level not seen since the Cuban missile crisis in the 1960s. The Soviet military hierarchy, in particular, was deeply suspicious of the intentions of the

---

[*] By then the plane had crossed the international date line, so the date was September 1, 1983.

United States and its president, Ronald Reagan, who had installed a new missile system in Europe and called the Soviet Union an "evil empire."

Some top Soviet officials were openly fearful that the United States was planning a preemptive nuclear strike against the Soviet Union. The Soviet leader, Yuri Andropov, was said to be consumed by such fear. In secret, the Soviet military had initiated an intelligence-gathering program to detect a potential nuclear attack. They had also ringed the country with a series of ground-based radars to aid their satellite system in detecting incoming warheads.

Less than a month after the KAL incident, the forty-four-year-old lieutenant colonel Stanislav Petrov was the duty officer serving the graveyard shift in the secret command bunker where the Soviets monitored their early warning systems. His training had been rigorous, and his job was clear: to validate any warnings the system might generate and report them to the senior military command. But unlike his colleagues, Petrov wasn't a professional soldier; he'd been trained as an engineer.

That night, Petrov had been on duty a few hours when alarm bells began to sound. An electronic map flashed. A backlit screen displayed the word "LAUNCH." Petrov's heart raced, and he could feel the adrenaline rush. He was in a state of shock. Soon the system reported another launch. Then another, and another, and another. The United States, the system was telling him, had launched five Minuteman intercontinental ballistic missiles.

Petrov's protocol stated clearly that the decision regarding whether to report any alarm was to be based solely on the computer readouts. Petrov checked the computer, and it rated the level of reliability of the alerts as "highest." The data on which the alarm was triggered had passed through thirty layers of verification. Petrov's job was now to simply reach for the phone and report the launches to the Soviet Union's top commanders, to whom he had a direct line. Such a report, Petrov knew, would almost certainly trigger an immediate and massive retaliatory strike. It would be the start of nuclear war. Petrov felt immense fear. There was some chance, perhaps minute, that this was a false alarm, and yet his report would be the end of civilization as we know it. But not to report it would be a dereliction of duty.

Petrov hesitated. The data reported by the computer was unequivocal, as were his orders. But something inside him made him focus instead on the possibility that it was a false alarm. He thought about it. He had no idea how, despite all the safeguards, such a grievous error could occur. Petrov realized he was running out of time. He had to take some action, one way or the other. The stress was enormous. He knew that a simple logical analysis, based on his orders and the data at hand, would dictate report-ing the apparent attack. But though he had no evidence that the alarms weren't real, he decided not to alert his superiors. Instead, acting out of an emotional aversion to starting World War III, he called the duty officer at the Soviet army's headquarters and reported a system malfunction.

Petrov knew that none of his professional-soldier colleagues would have disobeyed their orders, but he did disobey. Then he waited. If he was wrong, he'd be the biggest traitor in his nation's history, allowing the unanswered destruction of his country. But if so, would that really matter? As the minutes ticked by, he rated his chances at fifty-fifty. It wasn't until twenty minutes had passed, he later said, that he breathed easier. A later investigation would show that the false alarms were caused when, due to an unlikely alignment of sunlight on the tops of high-altitude clouds above North Dakota, Soviet satellites mistook the sun's reflection as multiple missile launches.

Emotions help us sort out the meaning of the circumstances we encounter. Especially in complex and ambiguous situations— and those in which we must make a quick decision—emotions act as internal guides that point us in the right direction. Though it may seem to have come from nowhere, Petrov's decision was the product of emotion that drew, in an instant, on the sum of his past experience in a manner that was both fast and difficult to match through rational analysis. What Petrov did that the more disciplined fighter pilot who shot down the KAL flight didn't do was let his emotions lead him.

Matters of the heart are the most important matters, and the most difficult to decipher. The new science of emotion has expanded our self-knowledge. We now know that emotion is profoundly integrated into the neural circuits of our brains, inseparable from our circuits for "rational" thought. We could live without the ability to reason, but we would be completely dysfunctional if we couldn't feel. Emotion is a part of the mental machinery we share with all higher animals, but even more than rationality its role in our behavior is what sets us apart from them.

2

The Purpose of Emotion

I was on the road, at a hotel, and wanted a beer. I called down for late-night room service. I was told to expect my order in about forty-five minutes. I didn't want to wait up that long. Because my order was simple, I asked, "Any way you could rush this?" The answer was "Sorry, no." A couple nights later, I found myself in the same situation. This time I tried a different tactic. "Any way you could rush this," I asked, "because I'd like to get it sooner than that?" This time the answer was "Sure, I can expedite it. I'll send it right up." An anecdote proves nothing of course, but my experience illustrates an effect that *has* been scientifically studied: routine requests are more likely to be granted when the requester provides a reason, no matter how obvious or flimsy.[1] That's because the person on the other end usually gives little or no thought to the reason: it's not the reason that triggers the cooperation, but merely the fact that one was given. Psychologists call that type of "mindless" reaction "reflexive." By that they mean that the link from the stimulus to the response satisfies three criteria: it must be triggered by a specific event or situation; it must result in a specific behavior; and it must happen virtually every time the stimulus is encountered.

The most famous reflexive reaction is the *knee-jerk reflex*, which is triggered when your doctor taps your knee tendon while it is in a relaxed state. Your response depends on that precise trigger: you won't jerk your knee in response to watching a video of a doctor wielding a hammer or after being startled by a door slamming. Conversely, your reaction is specific. When your knee is tapped, you don't shake your head or jump out of the chair; you simply move your knee. And, finally, your response is foreseeable: you do it virtually every time; in fact, it is very difficult not to. Such reflexes are necessary because if you had to actually think about all your movements, you could never move. Consider walking: it is governed by all kinds of reflexes that you do not think about (including the knee-jerk reflex), and all your brain has to do is give a general command to the spinal cortex to make many muscles all work together.

Physical reflexes such as the knee-jerk reaction don't require a mind; one can cut away an organism's entire brain, but if the spinal cord is left intact, it will still exhibit the knee-jerk reaction. But we also have more sophisticated reflexive reactions. One type is the *fixed action pattern*, or *script*, a little program our brains may follow when encountering familiar situations. The "autopilot" mode you may go into when driving to work, or mindlessly eating while pondering some problem or participating in a business meeting, also fits into that category. So does much of animal behavior, even that which may appear loving or thoughtful. For example, when a baby bird opens its mouth, the mother will fill it with worms or bugs. But that behavior has nothing to do with the bird being *her* baby, or *a* baby, or even a bird. It is a script triggered by anything that looks like a big gaping mouth; there's even a video on YouTube of a cardinal hopping over to feed a goldfish that's holding its mouth open.[2]

A more complex mental reflex is the psychological "button," the often-intense reaction we may have to certain social encounters. Just as your knee jerks when your knee tendon is tapped, a psychological "button" can be pushed when a triggering experience draws you back into some unhealed issues from your past. Some common triggers are if someone ignores you, or doesn't follow the rules, or lies to you, or criticizes you, or uses a phrase such as "you never" or "you always." Whether or not emotion was involved in the formation of the trigger/response cycle, if such an event produces an immediate unthinking reaction, that's the mental equivalent of the knee-jerk reflex.

Clinical psychologists run into this all the time; buttons can create havoc when they are pushed by our colleagues, friends, and family. Even in otherwise healthy relationships, they can result in repeated cycles of conflict. Once we learn our friends' and family members' buttons, it is wise to avoid pressing them. Once we realize we have buttons ourselves, it is wise to work at deactivating them. A friend who worked at home, for example, told me she used to bark at her husband when he'd walk into her office while she was focused on something. Once she realized this was a button—she'd been granted little privacy or respect for her personal space as a child—the occasional intrusions didn't bother her as much, and she was able to speak calmly to her spouse about minimizing them. Sometimes the remedy is as easy as learning to notice a button has been pushed and then using our conscious will to alter our reaction. We do something analogous when, while driving somewhere on autopilot, we switch to conscious control and alter our route to avoid a spot of bad traffic that we can see ahead.

One might be tempted to dismiss reflexive reactions as primitive and unimportant, but they can be powerful and are an important mode of operation in both nonhuman animals and ourselves. In simple organisms, they play a dominant role.

The power of reflexive behavior is exemplified in the success of one of the simplest organisms, the bacterium. Just as we humans seek to earn our living without working countless hours, these biological machines seek to maximize the food energy obtained per unit of foraging time. They accomplish that employing purely scripted "behavior." Through complex but automatic chemical means, they approach and devour nutrients and are repelled by encounters with noxious substances.[3] Bacteria even cooperate in groups, signaling one another through the release of certain molecules.[4]

"The variety of possible bacterial 'conduct' is remarkable," wrote the neuroscientist Antonio Damasio.[5] They cooperate with each other and avoid ("snub," some researchers call it) individuals that don't play along. Damasio tells of an experiment in which several populations of bacteria had to compete for resources within the flasks that housed them. Some responded with what looks like aggression, battling each other and sustaining heavy losses. Others survived by getting along. And all this went on for thousands of generations. If we humans have our Spartas and Nazi Germanys, and our pacifist states as well, so, too, do *E. coli.*

Though we humans have outgrown the kind of life that can be governed predominantly through reflexive reaction, such reactions govern more of our behavior than most people realize. For example, consider a pair of similar studies involving student volunteers who asked passersby for spare change.[6] One was conducted in a shopping area in San Francisco, and the other outdoors at a wharf in Santa Cruz. The volunteer panhandlers in both were students who wore typical school clothes such as T-shirts and jeans and made their appeals while maintaining a distance of at least three feet. To half of the passersby they approached—the control group—the students asked for either twenty-five or fifty cents. Those two appeals were about equally successful, bringing in money 17 percent of the time while also occasionally eliciting insult retorts like "Get a job" or "Panhandling is illegal here. We have a jail you might enjoy." But the vast majority of people just kept walking. Panhandling was common in those areas, and the researchers suspected that few of the passersby gave the request any thought. Instead, the scientists believed, most had responded automatically, employing a mental rule like "If a panhandler asks you for money, ignore the request."

The researchers hypothesized that they might increase the panhandlers' success if they could disrupt the script and cause the passersby to give the appeal mindful consideration. And so, to the other half of the passersby, they had the student volunteers employ a novel request—"Hey, buddy, can you spare thirty-seven cents?" That's about halfway between the twenty-five and fifty cents asked of the original group. The idea was that the unusual number would command people's attention, causing them to abort the application of the mental rule and instead deliberately consider the request. It worked, increasing the fraction who gave the panhandlers money in the San Francisco study from 17 percent to 73 percent. The strategy of increasing compliance in situations in which people typically pay little attention has been dubbed the pique technique: you may occasionally run into it, as with traffic signs I've seen once or twice proclaiming an oddball speed limit such as thirty-three miles per hour or a store item advertised as 17.5 percent off.

And that brings us back to emotion: reflexive behavior is a fundamental aspect of our evolutionary heritage, but at some point nature upgraded the approach to create an additional system for reacting to environmental challenges—one that is more flexible and hence more powerful. That's emotion.

Emotion is the next level up in our mind's information processing. As we'll see, it is far superior to strict, rule-based reflexive response. It enables even organisms with primitive brains to adjust their mental states according to the circumstances. That allows the correspondence between a stimulus and the organism's response to vary depending on the specific elements that are present in the environment or to even be delayed. In humans, the flexibility afforded by emotion also allows for input from our rational minds, leading to better decisions and more sophisticated actions.

THE ADVANTAGE OF EMOTIONS

Modern science has not always recognized the need for emotion, or its advantages over reflexive behavior. In fact, less than half a century ago scientists like the cognitive psychologist Allen Newell and the economist Herbert Simon (who would go on to win the Nobel Prize for other work) were still suggesting that human thought is fundamentally reflexive. In 1972, Newell and Simon presented volunteers with a series of logic, chess, and algebra puzzles and asked them to think aloud as they worked to solve them.[7] They recorded the sessions and then painstakingly ana-lyzed the participants' moment-by-moment reports, looking for regularities. Their aim was to characterize the rules of their sub-jects' thought processes in order to create a mathematical model of human thinking. Through that, they hoped to gain insight into the human mind and to discover a way to create "intelligent" computer programs that went beyond the limits of linear steps of logic.

Newell and Simon believed that the act of human reason-ing—of human thinking—was actually nothing more than a com-plex system of reflexive reactions. To be precise, they believed that thought could be modeled by what is called a production rule system. That's a collection of rigid if-then rules, which as a group result in reflexive reactions. For example, one such rule in chess would be "If your king is in check, move it." Production rules shed light on the way we make some decisions, and hence on some of our actions—for example, when people more or less mindlessly employ rules such as "If a panhandler asks you for change, ignore him." If human thought was really just a big system of production rules, then there would be little difference between us and a com-puter that runs an algorithmic program. But Newell and Simon were wrong, and their effort failed.

Understanding the cause of their failure sheds light on the purpose and function of our emotional system. Consider how a set of production rules can form a complete action strategy for a sim-ple system. Suppose it's below freezing outside and you want to program a thermostat to maintain the indoor temperature within a certain range, say, between seventy and seventy-two degrees. That can be accomplished employing the rules:

RULE 1: If temperature < seventy degrees, turn on heat.
RULE 2: If temperature > seventy-two degrees, turn off heat.

Whether you have a rickety old heater or a smart modern one, it is rules like these that form the basis of the heater's brain.

Such conditional commands form a primitive production rule

system; larger sets of rules are capable of governing more complex tasks. It takes about a dozen rules, for example, to represent the way grade school children subtract numbers, rules such as "If the lower digit is larger than the upper, borrow a one from the digit to the left of the upper digit." Some complicated applications can require thousands of such rules. They can be used to construct what computer scientists call "expert systems," programs meant to emulate human decision making in specific applications such as medical diagnosis and mortgage underwriting. In that application their approach had (limited) success. But production rules did not prove to be an adequate model of human thought.

The essence of Newell and Simon's failure lies in the richness of human life: though simple organisms such as *E. coli* can live by a set of reflexive rules, creatures with more complex lives cannot.

Consider, for example, what is involved in the seemingly simple task of avoiding tainted or poisonous food. Some such foods can be identified through their smell, and there are a great many types of such "bad" smells. Other tainted foods signal their unsuitability through appearance, taste, or feel, which also takes many forms. Sour milk looks and smells quite different from moldy bread. The degree of such indicators is also important. You may want to eat foods that look a bit suspicious but smell fine, depending on the prospects and challenges of finding alternative food. Or, if the food looks extremely odd, you might avoid it even if it smells fine. Or you might eat it despite its appearance if your body has been starved of nourishment. To employ a set of concrete, narrowly defined and rigid rules for every possible situation/response pair would overwhelm an animal's brain, and so another approach was needed.

Emotions provided that other approach. In a reflexive scheme, a specific trigger (for example, the milk smells slightly sour but I haven't eaten in days and there may not be other food and water nearby) produces a tailor-made and automatic response (for instance, drink it). Emotions function differently. The triggers are more general (the liquid looks and/or smells funny), and their direct product is not an action but a degree of emotion (mild disgust). Your brain then considers that emotion, along with other factors (I haven't eaten in days; there may not be other food and water nearby), and *calculates* your response. That eliminates the need for an enormous catalog of fixed trigger/response rules. It also allows far greater flexibility: you may consider a variety of responses (including to simply do nothing) and then make a considered decision.

In determining your response to an emotion, your brain takes into account multiple factors, in this case your degree of hunger, your aversion to having to venture out in search of other food, and other circumstances. That's where your rational mind enters the picture: once an emotion is triggered, our behavior results from a mental calculation based on facts, goals, and reason as well as emotional factors. For complex situations, it is this combination of emotion and rationality that provides the more efficient route to achieving a workable answer.

In higher animals, emotions also play another important role: they allow for a *delay* between the event that triggers the emotion and the response. That enables us to employ our rational thought to strategically temper our instinctive reaction to an event, or to delay it, waiting for a more opportune time. For example, suppose your body needs nourishment. You see a bag of Doritos. A reflexive response would be to mindlessly gobble them up. But because evolution has inserted an extra step in that process, when your body requires nourishment, you don't automatically eat any food in sight. Instead, you feel the emotion of hunger.* That nudges you toward eating, but now your response to the situation is not automatic. You might ponder the situation and decide to pass up the Doritos so you'll have room for the double bacon cheeseburger you're having for dinner.

Or think of what you're doing when the cable company representative has been extraordinarily uncooperative when you call up about a service issue. If humans operated reflexively, you might

---

* In modern research, hunger, along with thirst and pain, is called a homeostatic or primordial emotion.

lash out and say something like "go to hell, you idiot." Instead, the representative's behavior causes you to feel an emotion, such as anger or frustration. That emotion colors the way your mind processes the situation, but it allows input from your rational self. You might still lash out, but that's not automatic. You might instead ignore that impulse and say, after taking a deep breath, "I understand your policy, but let me tell you why it doesn't apply here."

Emotions can work that way in nonhuman animals as well, especially primates. Consider the book *Chimpanzee Politics: Power and Sex Among Apes* by the ethologist Frans de Waal. It's a lurid book, if you're a chimp. In it, de Waal describes how a young male who is aroused by a receptive female will wait and then, with her cooperation, find a way to mate out of view of the dominant males, who might punish him for it.[8] An alpha male, on the other hand, who, while on his rounds grooming supporters, receives a challenge from a younger male might ignore it and then launch an aggressive retaliation the next day. And a mother, upon having her infant taken by a juvenile, will stalk her until she gets an opportunity to snatch her baby back without risk of injuring the child.

Says David Anderson, a professor at the California Institute of Technology and member of the National Academy of Sciences, "In a reflexive action, from a very specific stimulus, you get a specific response, and you get it right away. That's fine if those are the only stimuli you encounter, and those are the only responses you need. But at some point in evolution, organisms required more flexibility, and the building blocks of emotion evolved to provide that."[9]

DO FRUIT FLIES CRY?

That Anderson is interested in the role of emotions, not just in humans, but also in more evolutionarily primitive organisms, is not surprising. His first research project, done while still an undergraduate in the 1970s, was on the molecular signals that come into play when a scallop comes into conflict with its nemesis, the starfish.[10] For Anderson, the key to understanding emotion lies in such research. He seeks to explain why biological information processors (that is, organisms) evolved the capacity for emotion, and how emotion factors into their processing (that is, "thinking").

Many people note that their dog or cat seems to have emotions, but what about simpler animals? "When I tell them about my work on that," he says, "they tend to think I'm crazy." He raised an eyebrow when he said it as if he were inviting me to speculate on that topic myself. I didn't think him crazy, but it also wasn't immediately obvious to me that his work wasn't. Anderson does research on emotion in fruit flies.

I questioned how much you could learn about human emotion by studying tiny creatures prone to kamikaze dives into my wineglass. Anderson chuckled—yes, fruit flies, like many humans, enjoy wine, and sometimes pay with their lives. Somehow, that started us talking about bars. I told him of a time recently when, late at night, I was walking the streets of Manhattan, and attracted by the music, I wandered into one. As I entered, what struck me most was the very large number of occupants, all roughly college age. Also, the music, loud on the outside, was *unpleasantly* loud inside. "It's bad for your ears," I said to the giant bouncer. He sneered and said, "If you were gonna lose your hearing, at your age, wouldn't that have happened already?"

I left, but afterward I told my son Nicolai about the scene. It's pretty standard, he told me. You go with a friend or two, get a drink, and talk as you scan the room. When a target has been identified, you walk over and chat her or him up. If after you exchange a few words there seems to be a connection, you graduate to the dance floor, where you try out the physical. If that works out, you leave together and mate (though he used a different term). Or, sometimes you don't. Sometimes the person already has a significant other. "What happens then?" I asked. "You feel rejected and go have a drink," he told me.

The specifics of this ritual are a mix of old and new and through the ages have been driven by human emotions like lust and love. Can we really learn anything about such complex human passions,

I asked Anderson, by studying fruit flies? Apparently, I had played right into his hand: fruit flies, it turns out, follow a mating ritual that has surprising similarities to that of Nicolai and his friends.

In the fruit fly world, the male initiates the ritual by approaching the female. There are no pickup lines, of course. Instead, he taps her with his foreleg. There is also music; he generates it by vibrating his wing.[11] If the female accepts the advance, she will do nothing, and the male will take over from there. But not all female fruit flies are receptive; if a female already has a boyfriend—that is, if she has mated with a different male—she'll turn down the advance. She does that either by striking him with her wings or legs or by running away. And now the punch line: as I said, fruit flies favor alcohol, and if a male is rejected, and a source of alcohol is available, the male will be likely to respond, like Nicolai, by having a drink.[12]

So fruit flies have a lot in common with Nicolai. But are they, like Nicolai, driven by emotion? Or are the flies acting reflexively following fixed scripts that encode their mating behavior. And how can one perform experiments to determine which it is? Anderson's goal wasn't to investigate whether all animals exhibit emotion or show that no animal behavior is reflexive (as I've said, even humans at times behave reflexively); he was simply interested in whether emotion can play an important role, even in "lower" animals.

Those are difficult questions, for emotion scientists have no good, or even generally accepted, definition of the term "emotion." In fact, one group of researchers wrote an article that did no more than categorize the various distinct definitions that emotion researchers employed.[13] They found ninety-two. And so, with his Caltech colleague Ralph Adolphs, Anderson decided to embark on a modern exploration of the defining traits of emotion—across the animal world—a kind of update to Darwin's pioneering work. They identified five features that are most salient: valence, persistence, generalizability, scalability, and automaticity.

Imagine an ancient ancestor walking on the African savanna. She hears a snake and jumps out of the way. If reflexive responses guided all aspects of life, that ancient woman would then continue walking without taking into account that the presence of one snake might indicate a raised probability of encountering others.

Thanks to emotions, our reactions and those of other animals, even fruit flies and bees, can be more sophisticated than that. If you're hiking and you hear a snake, your heart continues to pound for at least several minutes after you jump out of the way. During that time even a rodent rustling in the underbrush may cause you to jump. That illustrates the first two properties of emotion that Anderson and Adolphs identified: *valence* and *persistence*.

Emotions have value of some kind: they can be positive or negative, lead to approach or withdrawal, to feeling good or feeling bad. In this case, you jumped away. That's withdrawal, or negative valence. Persistence refers to the fact that after you jump away from the snake, your fear response does not immediately disappear. It lingers, leaving you in a hypervigilant state. Because misidentifying a rodent as a snake has little negative consequence but reacting too slowly to other lurking snakes could be deadly, the persistence of emotion was useful in helping our ancestors detect and avoid environmental dangers. A more modern example comes from my friend June, who'd spent a frustrating and maddening hour online trying to straighten out a serious computer problem. Shortly after getting it resolved, her ten-year-old, playing with a basketball indoors, knocked over and broke a vase. Because her negative feelings had not immediately dissipated, what might have been a simple scolding became a screaming episode.

Anderson and Adolphs's third salient quality of emotion is *generalizability*. In a reflexive reaction, precisely defined stimuli lead to particular responses. To say that emotion states are generalizable means that a whole variety of stimuli may lead to the same response and, conversely, that we may at different times exhibit a variety of responses to the same stimulus.

A primitive laboratory jellyfish, when poked, will always crumple up and sink to the bottom of the dish. That's reflexive behavior. The jellyfish won't, before reacting, stop to consider who did the poking, or why, or whether now is a good time to be sitting at the dish's bottom. Julie, on the other hand, when unfairly criticized by her boss might react in different ways. She might withdraw, or she might "push back." The response she exhibits depends not just on the triggering event but also on various other factors that her brain takes into account when calculating her reaction. How well has she been performing at her job lately? What is her boss's mood today? How has their relationship been going?

Scalability is the fourth quality that distinguishes emotion states from mere reflexive behavior. In a reflexive reaction, after the stimulus occurs, you exhibit a fixed response. Emotion states and the responses that are produced by them, on the other hand, can scale in intensity.

Depending on what else is going on in your life, or at the moment, a particular incident might make you feel a little sad, causing the corners of your mouth to turn down. Or it might make you very sad, and your eyes might start tearing. Emotion states allow for a gradient of intensity of reaction to the same stimulus, depending on a spectrum of other relevant factors. A strange noise coming from downstairs when you'd thought you were alone might make you a little fearful if the incident happens at noon, but very afraid if it's at midnight. The difference in reaction is a useful distinction based on your knowledge of the world (specifically in this case on your knowledge of when home break-ins are most likely to occur). It is made possible by the scalability of emotion, and not characteristic of the one-size-fits-all approach of reflexive processing.

Finally, Anderson and Adolphs say emotions are *automatic*. By that, they do not mean that you have no control over the emotion. They mean that, like reflexes, emotions arise without intention or effort on your part. However, although they arise automatically, unlike reflexes, emotions don't cause an automatic response.

When someone cuts in front of you in line, your anger will

come automatically, but because you don't want to make a scene (or because the person is a lot bigger than you), you might work at not expressing your anger. If you are at a dinner party and eat what you suddenly realize is kidneys and you hate all organ meats, the disgust is automatic, but you may make an effort not to gag in order to avoid offending your host. That kind of control over emotions is most pronounced in adult humans. Human children have far less control, because that ability is linked to the maturation of the brain. Which is why it takes a while to train them not to just spit out whatever they don't like.

EXPERIMENTING ON EMOTION IN THE LAB

One of the nice aspects of the Anderson-Adolphs characterization of emotion is that each of the features they identified can be tested for in the lab, even in primitive animals. That brings us back to the fruit fly. In a series of clever experiments Anderson and others were able to show that fruit flies in various situations are reacting on the basis of emotional states—as characterized by valence, persistence, generalizability, scalability, and automaticity—and not just reflexively.

For example, fruit flies become startled by certain events, such as the sudden appearance of a shadow, or when they feel a puff of air, both of which presumably indicate that a predator might be nearby. Is that a reflex, or are the fruit flies truly in a state of fear? To investigate, the scientists created an environment in which they could startle the fruit flies while they were feeding. That presented the fruit fly with an important choice. If the fly runs or flies away and there was no predator present, it wasted time and energy and has to return later to compensate for the burned calories. But if it doesn't race away and a predator is present, it could get eaten.

Anderson found that when first presented with a shadow, the flies would jump off the food and then return after some seconds had passed. If they were then presented with a second shadow, they would exhibit a modified response: they would jump off again, but this time stay away longer. Because the trigger—the shadow—was the same in both cases but the response was different, this was not a reflexive reaction.

What's more, the fruit fly's response clearly exhibited valence, because it sought to avoid the shadow. It also exhibited both persistence and scalability: the first incident put the fruit fly in a state of fear, which persisted, and was then scaled up by the second appearance of the threat.

Such nuanced emotion-based responses are more effective and efficient than simple reflexive behavior, which might dictate jumping away at the sight of a shadow and staying away for a prescribed period but not take into account the increased probability of danger implied by the fact that the shadow appears repeatedly.

The fruit flies that showed a preference for alcohol after experiencing sexual rejection also seemed to be experiencing a persistent emotion—rejection—and were seeking to reset that negative emotion state by ingesting the ethanol, which experiments have shown is rewarding to them (they will perform tasks in order to achieve access to it).[14] As in humans, individual fruit flies vary in the degree to which their emotion states exhibit the above qualities. As has been shown in the research on emotional intelligence, having an awareness of the dynamics of our emotion states is an important component of success in life. It helps us to motivate ourselves, to control impulses and regulate our moods, and to respond appropriately to others.

A fruit fly brain has a hundred thousand neurons (half in the visual system) compared with the roughly hundred billion in a human brain. That's just one-millionth the number, but a fruit fly can perform amazing aerodynamic maneuvering: it can walk, it can learn, it follows courtship rituals, and most impressive of all, it displays fear and aggression—an indication of the essential role emotions play in the information processing of all animals.

Our own emotional minds appeared long after the fruit flies', which came about forty million years ago. But our evolution for the most part occurred long before our settling into towns and cities. That means that though our emotions evolved to help our brains calculate our reactions, the very traits that made them useful hundreds of thousands of years ago can also cause behavior that is inappropriate to our current civilized existence. Generalizability may result in your exhibiting a response fit for fending off a predator but not for dealing with a driver who cuts you off in traffic. Scalability allows you to ratchet up the intensity of your reaction, but also means that at times you may "lose it." Persistence may leave you in a hypervigilant state all day, overreacting to events that occur long after you've forgotten the initial incident that put your mind in the state of alert.

As a child, I watched scientists study animals on the old *National Geographic* program. Before I'd ever had sex, I'd seen detailed footage of a praying mantis couple copulating. During their mating, the female bit her partner's head off. For a prepubescent, that was too much information. I wondered if there was some hidden metaphor. But the truth is, there weren't many studies of human sexuality or even of human emotion back then. We seemed to understand much more about the behavior of animals than we did about humans. Those were the days when, even among psychologists, it was commonly thought that emotion was to be avoided, even the emotion of maternal love. Said one child-rearing manual, "While nature has thus wisely endowed the mother with all-embracing love for her children, it would have been better if nature had equipped the mother so that she could control her affection by her reason."[15]

Affective neuroscience teaches us a different lesson, that emotion is a gift. It helps us to quickly and efficiently make sense of our circumstances so that we can react as necessary; it feeds into our rational thought, allowing us, in most cases, to make better decisions; and it helps us connect with and communicate with others. To understand the purpose and function of emotions doesn't diminish the role they play in making our lives richer; it enables us to better understand what it means to be human.

3

The Mind-Body Connection

Simon was one of the leaders of the anti-Nazi underground in Czestochowa, Poland. The Jewish ghetto in which he lived was surrounded by walls and fences, sealing the neighborhood and, in all likelihood, the fate of its inhabitants. Still, these fighters did their best to mount a resistance.

On occasion, after darkness had enveloped the city, some would sneak out to procure goods and to commit acts of sabotage and theft. One such night, Simon and three cohorts crept to a barbed-wire fence that ran along the dirt in a quiet, isolated area. There, they dug until they could pull up the bottom of the fence, enabling them to squirm under it to the other side. Simon held the fence taut as the others crawled through. Then he took his turn.

A hundred meters away, a German soldier was waiting in a small truck. He'd been paid off to take them to that night's destination. As Simon's fellow fighters crept toward the truck, Simon pushed under the fence to join them. His clothes caught in one of the sharp, unfinished ends. By the time he'd worked himself free, the others were on the truck, and the impatient driver had begun to pull away.

Simon was left with a choice, and only a moment to decide. If he ran to the truck, he could probably catch up with it. But that would risk drawing attention to them and getting them all killed.

If he let them go on without him, they'd have to face their mission with one less comrade than planned, also a dangerous option. Neither action nor inaction appealed to him. But as the truck rolled forward, Simon knew that hesitation was equivalent to the choice of staying behind. He quickly weighed the pros and cons and decided to make a run for the truck.

As he began to take his first step, he suddenly paused. He didn't know what stopped him. It wasn't fear, he told me. He'd gone on similar missions many times—danger had become routine—and as snafus went, this was a relatively minor one. Yet his body seemed to be reacting to something. The Germans made them live like animals. Had his animal self taken over? Had his eyes and ears registered a suspicious pattern that was too subtle to rise to his conscious awareness? He never discovered what his body was telling him, but in the end the impulse to freeze shaped his actions: he knelt to the ground and watched as the truck continued down the road.

The truck hadn't gone far when a vehicle packed with SS—Hitler's genocidal paramilitary organization—came seemingly out of nowhere and raced toward it. The SS officers intercepted the truck. A moment later, they gunned down its occupants. Had Simon not been held up by his primitive visceral reaction, he would have been killed with the others. Had that happened, I would not be writing this book, for a dozen years later Simon, then a war refugee living in Chicago, would have his second baby boy—me.

My father grew emotional as he told me this story, decades after the fact. He had come so close to being killed and struggled to make sense of his survival. He said he felt no fear. Yet he hesitated. What had saved him? What had made him decide to hang back when, in countless comparable situations, he had always pushed forward? It wasn't a conscious decision based on anything he was aware of having observed. To him, the situation had seemed routine. His rational mind had told him to chase after the truck and join his comrades. And yet his body knew different, and it held him back.

If you've ever been stumped by a challenge, puzzle, or problem and then had the answer pop into your mind while you were engaged in some unrelated activity such as jogging or taking a shower, you've experienced the fact that your unconscious mind can process information "in the background," so that you are not even aware of it. Today we know that when your body is in a state of high alert, your unconscious mind engages in a similar problem-solving activity, this one with the goal of keeping you safe. Your unconscious brain becomes hyperaware of your bodily state and the threats in your surroundings, and then it goes to work calculating whether your survival is endangered and, if so, what you should do about it. From that interaction of mind, body, and sense comes an intuition or impulse, aimed at self-preservation.

That is what led my father to override his conscious will to join his fellows: while his conscious mind was pondering one set of facts and goals, his unconscious had analyzed additional information, subtle clues about his environment and his bodily state that had not pushed their way into his consciousness. The origin of that primal awareness of danger is a kind of built-in sensor our brains have that monitors our bodily condition and threats from the environment. To describe that sensor system, the psychologist James Russell coined the term "core affect."*

CORE AFFECT

Core affect is a reflection of your physical viability, a kind of thermometer whose reading reflects your general sense of well-being, based on data about your bodily systems, information about external events, and your thoughts about the state of the world. Like emotion, core affect is a mental state. It is more primitive than emotion, and it emerged much earlier in the evolutionary time line. But it influences the development of your emotional experience, providing a connection between emotion and body state. The connection between core affect and emotion is still

---

* The word "affect," in "core affect," rhymes with "aspect."

not well understood, but scientists believe it is one of the most important factors or ingredients from which your emotions are constructed.

Whereas emotion has the five key characteristics delineated by Anderson and Adolphs—and can take a number of specific forms such as sadness, happiness, anger, fear, disgust, and pride—there are only two aspects to core affect. One is valence, which is either positive or negative and describes the state of your well-being; the other is arousal, referring to the intensity of the valence, how strongly positive or negative it is. Positive core affect means your body seems to be doing well; negative core affect sounds the alarm, and if the arousal is high, that's a loud, urgent alarm that is difficult to ignore.

Though principally a reflection of your internal condition, your core affect is also influenced by your physical environment. It responds to art and entertainment, to funny or tragic scenes in a film. And it is affected directly by medicines and chemicals, both uppers and downers and euphoric drugs. In fact, the core-affect-altering properties of many drugs are precisely the reason that many people take them—uppers to increase arousal, downers to decrease it, other drugs, from alcohol to Ecstasy, to help induce feelings of positivity.

Your core affect is always present, just as your body always has a temperature, but you are only consciously aware of it when you focus on it, such as when someone asks you how you are doing or when you pause to ponder that yourself. Core affect sometimes varies noticeably from moment to moment, but it can also be more or less constant over long periods of time. As a conscious experience, psychologists describe valence as the degree of pleasantness or unpleasantness you feel at a given time. It is what you experience when you feel cheerful because you're healthy and your day is going well and you've eaten a good meal, or miserable because you have a bad cold and you're hungry.

Arousal as a conscious experience is characterized by the degree of energy you feel: energetic at one end of the spectrum, maybe because you're listening to music that stirs you or taking part in a political demonstration that excites you; sleepy or lethargic at the other end of the spectrum, perhaps because you're listening to a classroom lecture that bores you (though I can't imagine that occurred when *I* was teaching).

In the genesis of an emotion, core affect is believed to represent the input of your body, which, when combined with the circumstances you find yourself in, the context of that situation, and your background knowledge, will produce the emotions you experience. One can think of it as a sort of baseline state that can influence your emotions in any specific situation and the decisions you make as a result—decisions often attributed to intuition—such as my father's decision to stay behind. It is thus a crucial link between body and mind, connecting your physical condition to your thoughts, feelings, and decisions.

If you win $10,000 in the lottery, you'll probably be in a happy mood all day and for many days that follow. Meanwhile, your core affect may also spike in both positive valence and degree of arousal; after all, the mountain of money is good news for your survival in general. But core affect is more tied to your physical than financial well-being, so despite the good news it will grow negative if you miss lunch and are hungry, it will diminish in degree of arousal as you grow tired, and it will immediately take a nosedive if you bang your head on a door frame, only to recover in a few minutes.

To understand how core affect works—and the significance of the mind-body connection—it helps to go back to the writings of the Nobel laureate physicist Erwin Schrödinger, who, writing in the 1940s, defined life as a battle against the law of entropy.

The law refers to the tendency in nature for physical systems to grow more disorderly in time. For example, if you release a drop of ink into a glass of water, it won't stay in that nice droplet form for long, but will soon become amorphous and spread throughout the glass. Most highly ordered objects in nature will eventually suffer that fate. But the tendency for entropy or disorder to grow is absolute only for isolated systems; it need not hold for objects that interact with their surroundings. Life-forms are such systems; they interact in part by consuming food and absorbing sunlight, and those actions allow them to overcome the law of entropy. A crystalline block of salt left out in the elements will eventually break apart or dissolve in the rain. A living thing will take action to resist its own destruction. That's the defining property of life, Schrödinger said: life is matter that actively counters nature's tendency toward increasing entropy.

The battle to maintain life is fought on many levels. The "atoms" of life are the cells that make up our bodies, and each individual cell executes processes that help stave off the increase of entropy. But a cell's success at that is not eternal. An encounter with too much heat or cold, or the wrong chemical, can disrupt a cell, causing it to dissipate, to cease its short existence as life, or, as the Bible says, to return from ashes to ashes, from dust to dust.

In a multicelled organism, the battle against disorder is also fought on a larger scale. In animals, the brain and/or nervous system acts to regulate organs and bodily processes and to maintain their function within certain parameters so that they work together seamlessly to maintain life. "Homeostasis," from the Greek words for "same" and "steady," refers to that ability of an organism, or an individual cell, to maintain its stable internal order even in the face of changes in the environment that could threaten it. The term was popularized by the physician Walter Cannon in his 1932 book, *The Wisdom of the Body*, which details how the human body maintains its temperature, and keeps other vital conditions such as the water, salt, sugar, protein, fat, calcium, and oxygen contents of the blood, within an acceptable range.[1]

Fighting off threats to homeostasis requires constant monitoring and adjustment. On the microscopic scale, cells sense their internal state and the external conditions and react according to fixed programming that evolved over aeons. As multicellular organisms evolved, each cell of the organism maintained such processes, but higher-level mechanisms such as core affect also evolved.

Core affect, in this context, is a neurological state in higher animals that acts as a sentinel to watch for threats to homeostasis and influences an organism to respond accordingly.[2] As I said, having just two dimensions—valence and arousal—core affect is unlike the nuanced state that we traditionally think of as emotion. And though your emotional experiences such as fear seem to arise from networks with nodes in many brain regions, core affect is correlated with activity in two particular regions.

Valence—pleasant or unpleasant, positive or negative, good or bad (or somewhere in between)—corresponds to the message "everything seems fine" or "something is wrong." It originates in the orbitofrontal cortex, a part of your prefrontal cortex that sits just above your eye sockets.[3] It is associated with decision making, impulse control, and the inhibition of behavioral response—all tasks that would have been important in my father's hesitation at the fence that night.

The arousal dimension of core affect represents neurophysiological alertness—the person's state of responsiveness to sensory stimuli. It is a measure of the magnitude of that responsiveness—strong or weak, energized or enervated. Arousal is correlated with activity in your amygdala, a small almond-shaped structure that is known to play a role in the generation of a number of emotions.[4]

That core affect is correlated with activity in the orbitofrontal cortex and amygdala is no accident. Those structures are known to be important in decision making and have extensive connections to sensory areas and regions of the brain involved in emotion and memory. They have continuous access to information about the state of your body and of your surroundings. By integrating that information, core affect reflects whether the homeostatic state of the body and its current external circumstance is conducive to survival and provides an appropriate undercurrent that colors all our experience and every action we take.

WHEN JUNCOS GAMBLE

The power of core affect was nicely illustrated by an experiment by Thomas Caraco, a biologist then at the University of Rochester, who studied the phenomenon in the 1980s, long before it became a matter of interest in psychology, before the term had even been coined.[5] For his research, Caraco captured four dark-eyed juncos, a species of small songbird, in upstate New York, kept them in separate aviaries, and conducted eighty-four experiments on them.

In one, the birds were given a choice between two trays of a food they like, millet seeds. In training sessions, the juncos had learned that one tray had a fixed number of seeds, while in the other the number of seeds varied, though on average the number equaled the number of seeds on the first tray. In the experimental trials, the two trays would be offered simultaneously at opposite ends of the aviary, equidistant from the perch of the hungry bird, and it would have to choose which to feed at. This mimics a trade-off often encountered in nature, and in our lives: take a sure thing, or gamble on getting something better, at the risk of getting something worse.

The trick in the experiment was that the birds were kept at different temperatures, and that variation in their bodily state affected their choice: when they were warm (corresponding to positive core affect), they preferred the fixed option, but when they were cold (negative core affect), they chose to gamble. That makes sense because when the juncos were warm, the fixed-seed option was sufficient to nourish them, so why take a risk? But when they were cold, they required more calories to maintain homeostasis, so only the second tray, though a gamble, offered them the chance at getting the calories they needed.

It's the kind of choice we make all the time in human society. Imagine that job A pays more than job B but offers less job security. If both jobs meet your income needs, you might be inclined to accept the more secure, lower-paying job. But if not, you could be more inclined to take a chance on the more lucrative job. It's doubtful that the juncos employ the same type of conscious reasoning that we do to make such decisions, but by monitoring their own bodily state and taking it into account in their instinctive mental calculations—that is, through the influence of their core affect—they came to the same conclusion that would have been reached by a professional employing the mathematics of risk analysis.

Though we humans have the power of logical thought, like juncos, our core affect primes us to think, act, and feel a certain way. We all react differently, at different times, to the same situation, and that difference in our response is often due to the hidden influence of core affect. Understanding the power of core affect is thus an important part of gaining perspective on how you react to others, and on how they treat you.

If, Saturday morning, after a good breakfast and a pleasant cup of coffee you receive a telemarketer's call, you might react politely. Your comfort level allows you to have a response that comes from sympathy for the plight of a person desperate enough to take such a job. On the other hand, if you wake up with a sore throat and a cough, you might curse the caller and slam the phone down, focused on your feeling of resentment at being interrupted on a weekend morning. Your behavior in both cases is as much a reflection of your own psychological state as it is a reaction to the event. In touchy situations in particular, it is good to keep in mind that a person's response to your words or deeds might be influenced as much by that person's current core affect as by anything you have said or done.

THE GUT-BRAIN AXIS

Communication of core affect to the mind occurs through neurons but also through the action of molecules circulating in the blood or distributed in the organs, such as the neurotransmitters serotonin and dopamine. Core affect is a central element in the mind-body connection, which we now know is far more powerful than scientists, even a decade or two ago, used to believe. The turnabout has been so strong that what used to be considered borderline "crackpot" ideas have now become mainstream. For example, consider the recent academic science embrace of meditation and mindfulness; although practitioners don't express it this way, both are routes to becoming aware of your core affect at all times.

The evolutionary roots of the mind-body connection go back to the beginning of life itself. Long before animals came along, before eyes and ears and noses evolved, organisms as primitive as bacteria could sense other organisms and molecules in their immediate neighborhood, and they could monitor their internal state. Evolution had not yet invented minds, but these early organisms reacted to that information when "choosing" what processes to execute.

John Donne wrote in 1624, "No man is an island entire of itself; every man is a piece of the continent, a part of the main."[6] The same holds for cells. As I mentioned earlier, even bacteria do not survive on their own, but rather exist in groups that signal each other by releasing certain molecules. In that manner, each cell's struggle against entropy is aided by the experience of its peers. It is this molecular signaling that allows bacteria to build up resistance so that they can survive antibiotics. Many of those drugs work by dissolving the bacterium's membrane. But before the bacterium dies, it may emit a molecular distress signal, causing other bacteria to engage in protective behaviors that alter its biochemistry. If not enough antibiotic is administered, the bacteria "learn" the avoidance behavior before they are all wiped out, and the disease is not cured. That's why your doctor always tells you not to stop taking the antibiotic until you have completed the whole prescribed course, even if you think you are well and no longer need it, because the disease can come roaring back, perhaps even stronger.

Bacteria were among the first life-forms, originating almost four billion years ago, but their ability to sense their own state, and that of the environment, and to emit signals so that other cells can adjust, is the underpinning of core affect. How does such a mechanism, suited to individual cells, evolve into a key process within the human body?

After bacteria, the first great leap toward higher animals occurred about 600 million years ago, when multicellular organisms evolved. These took the bacterial colonies to their logical extreme. The interacting colony became a single multicellular creature, and what had been communication between independent cells was now communication among cells of the organism. Eventually, different types of cells evolved within an organism, the analogue of the different tissues of the human body. Soon thereafter, nerve cells evolved, organized into what scientists call nets—simple sets of neurons that were connected in a diffuse network across an organism's body but not concentrated in a separate organ.

One of the main functions of the newly evolved nerve nets was to run digestion.[7] That's vividly illustrated in the hydra, a throwback to those ancient times that the neuroscientist Antonio Damasio calls "the ultimate floating gastronomical systems." Basically tubes that swim around, hydras open their mouth, execute peristalsis, digest what floats through, and shoot the rest out their other end. It is in the sensing and reacting executed by organisms such as these that we see the beginnings of core affect. We are far more complex than hydras, but our core affect system is essentially a "grown-up" version of the body surveillance capability that evolved in those creatures. In fact, when anatomists study the gut's nervous system—called the enteric nervous system—they find a striking resemblance to those ancient nerve nets.

A sophisticated system of nerves sometimes called our "second brain," the enteric nervous system regulates and runs throughout our gastrointestinal tract. It has only recently been studied in detail, but the "second brain" nickname is well deserved because the enteric nervous system can make its own "decisions" and operate independently of our brain. It even employs the same neurotransmitters. For example, 95 percent of our serotonin is in our gastrointestinal tract, not in our brain. But though our enteric nervous system can operate independently, it, and our entire gastrointestinal tract, are intimately connected to our brain and central nervous system. So the idea in popular culture that our gut is closely tied to our mental state has a strong basis in science.

The connection between gut and brain is so important it has

a scientific name: the gut-brain axis. It is through the gut-brain axis that our gastrointestinal system exerts its outsized influence on our core affect.

Our feeling of physical well-being, for example, is rarely affected by what's transpiring in our spleen, but it is often affected by the state of our digestion. Our core affect, in turn, influences our gut, forming a feedback loop: if you are in sudden danger and your core affect turns negative with high arousal, you might get heartburn, indigestion, or a "sinking feeling" in your stomach. Recent and intriguing research shows that there also seems to be a relation between bowel disorders and psychic disturbances such as chronic anxiety and depression.[8] It has long been known that a distressed brain can disturb colonic function, but the new research suggests that the causal arrow also points in the other direction: disruption of the gut contributes to neuropsychiatric disease. This seems to happen through complex biochemical processes; for example, alterations in the bacterial environment can degrade the intestinal barrier, allowing undesirable neuroactive compounds to gain access to the central nervous system.

From the point of view of evolution, the nerve nets that bear such resemblance to our second brain preceded by about 40 million years the development of actual brains, in which neural processing and sensing are physically separated from other cellular functions. Planaria, a kind of flatworm that can regenerate body parts, date back to the era, 560 million years ago, when the brain first evolved as a distinct organ. Although they have actual brains, there is so little differentiation between the planarian brain and body that if a planarian's brain is excised, the new brain it regenerates can retrieve the organism's old memories from its remaining bodily system.[9]

Another dramatic illustration of the mind-body connection— especially in digestion—comes from an astonishing experiment with mice.[10] In it, scientists separated two sets of mice, one set timid, the other adventurous. They then took the gut microbes from each set and transplanted them into another set of mice that had been raised to have a relatively sterile gut. Moving microbes from animal to animal may seem to be an odd exercise, but recent research shows that the microbes within the gut are so influential to its operation that transplanting microbes is like transplanting part of the gut itself. And that "partial gut transplant" had an astonishing effect: once the microbes had multiplied and colonized their new host, the recipient rodents took on the personality traits—either timid or adventurous—of the set of mice whose microbes they had received. What's more, other research suggests that transplanting fecal bacteria from humans with anxiety into mice can lead to anxiety-like behavior among these rodents, whereas transplanting bacteria from calmer control humans does not.[11]

What about humans? Scientists have now performed MRI scans to examine the brains of thousands of volunteers and compared their brain structure with the mix of bacteria living in their guts. They found that the connections between brain regions differed depending on the species of bacteria that dominate. The studies suggest that, as in the mice, the specific mix of microbes in our guts may influence how our brain circuits develop and how they're wired. More research is needed, but it seems that the influence of bacteria on the subjects' core affect may play an important role.

An enterprising medical researcher, on hearing all this, may wonder if a course of powerful antibiotics followed by the ingestion of gut fluid from another person could alter some unwanted personality trait. Would putting grim aunt Ida on penicillin for a week, then having her eat the vomit of a happy person, turn her into Mary Poppins? Maybe. In the past few years, scientists have investigated fecal transplants as a treatment for disorders such as chronic anxiety, depression, and schizophrenia.[12] The field is still in its infancy, but perhaps someday we'll have treatments of that sort. For now, what this research illustrates is that the separation of brain and body is artificial. They are a single, thoroughly integrated, organic unit, and core affect is an important part of the system.

WHY NOTHING GOOD COULD COME FROM
A HEAD TRANSPLANT

In the 1960s, Western culture was far from accepting the importance of the mind-body connection. If you search Google for use of the term "mind-body connection" (in quotes) over the past decade, several hundred thousand mentions come up. If you search over the decade 1961–1970, you get five hits. Two are in foreign languages. Of the others, one is an article on Jewish spirituality, another a court transcript of a gruesome murder.

Despite its avant-garde nature, some forward-thinking scientists of that day did investigate the idea. One was George W. Hohmann, a psychologist at the Veterans Administration hospital in Long Beach, California. Hohmann was a paraplegic who had sustained a spinal cord injury while serving in World War II.[13] Such injuries can cause an impaired ability to control or activate muscles, but the spinal cord also carries sensory signals, so victims of that malady may lack awareness of heat, cold, pressure, pain, the position of their limbs, and even their own heartbeat. At the VA, Hohmann was in daily contact with others who had spinal cord injuries. He wondered, if the state of the body is an important input to emotional feeling, then does the lack of bodily feedback diminish the intensity of the emotions these patients feel, as it seemed to do in his own case? To find out, he probed the issue in interviews with twenty-six male patients, asking them to compare certain of their emotional feelings before and after injury.[14] In a now-classic article, he concluded that paraplegic patients did seem to experience "significant decreases in experienced feelings" of anger, sexual excitement, and fear. In recent years, studies of the emotional responses of paraplegics have supported that finding.[15]

Today we know that the human brain-body connection is so vital that if one could sever the spinal cord and other nerves, and the blood vessels that connect a head to the lower body, and then as carefully sew that head back onto another headless body, the disruption of the brain-body feedback loop would be a major factor threatening the new organism's survival. That may seem like an unlikely and off-the-wall example, but there have been a number of attempts over the years to do just that. In fact, head transplants have had such a long and full history that a Harvard Medical School surgeon recently wrote an article in a surgical journal on the topic: "The History of Head Transplantation: A Review."[16]

The article begins by describing the first such effort, in a dog, performed more than a century ago by the surgeons Alexis Carrel and Charles Guthrie. The dog was able to see, make sounds, and move, but died after a few hours. For his work on transplantation, Carrel received the 1912 Nobel Prize in Physiology or Medicine. A Russian surgeon, Vladimir Demikhov, repeated the feat in 1954, creating a dog that survived for twenty-nine days—but this time, no Nobel Prize. In the ensuing years, such surgeries have been performed on mice, and even on primates. In 1970 a rhesus monkey with a transplanted head survived for eight days and was deemed, "by all measures, normal."

We all have different definitions of "normal," and as one who has had a few surgeries in my life, I know that when your surgeon promises you that after the operation you'll soon be "normal" again, you'd better ask for his or her definition of that word. I'd certainly hope that coming out of it as a severed head on a gurney doesn't qualify. What this surgeon meant was that the monkey was able to bite, chew, swallow, track with its eyes, and exhibit characteristic wakeful EEG patterns. That's about it. Meanwhile, it needed constant infusions of drugs and intermittent mechanical ventilation so that it wouldn't suffocate. There was no swinging from tree to tree and grabbing bananas for this, "by all measures, normal" monkey.

Given that state of affairs, no one would consider performing such an operation on a human, right? It turns out the topic is not so off the wall: in 2017, Sergio Canavero of Italy and his colleague Xiaoping Ren of China announced a plan to transplant a human head from a living person onto a fresh donor cadaver, presumably of someone who'd died of a head injury.[17] According to these doctors, the procedure was made possible by recent advances in immunotherapy to prevent rejection of the new head, and in the technology of deep hypothermia, allowing the head to survive for the time required to attach it to the target body. The plan is to clamp off and sever the neck arteries and veins and cut the spine and its nerves between the fourth and the sixth cerebral vertebrae and then reattach it all while a pump keeps the blood flowing, and at a temperature of eighty-four degrees.

Who would volunteer for such a gruesome experiment? The surgeons seemed confident they'd find plenty of prospects among the terminally ill. It's possible. They plan to do the surgery in China because no American or European institute would permit such an operation. But they are not "mad scientists," these doctors wrote in *Surgical Neurology International*. "Western bioethicists needed to stop patronizing the world," said Canavero.

There are of course many reasons that Canavero's proposed operation is a bad idea, even apart from purely ethical considerations. Not only because the surgery hasn't been well tested or particularly successful in lower animals, but also because it is projected to cost about $100 million, and the patient will almost certainly soon die, and may experience severe pain. All that aside, given the paramount importance of the mind-body connection, what would such surgery, if physically successful, do to the patient's core affect, emotional well-being, and psychology in general?

Ren and Canavero recognize that problem. They discuss the case of another transplant, whose failure they blame on poor integration of the body part into the patient's body image. "We recognize that accepting a foreign body part as one's own requires psychological resilience," they wrote. And that was a *hand* transplant.

Paul Root Wolpe, editor of the *American Journal of Bioethics Neuroscience*, wrote, "Our brains are constantly monitoring, responding to, and adapting to our bodies. An entirely new body would cause the brain to engage in a massive reorientation to all its new inputs, which could, over time, alter the fundamental nature and connective pathways of the brain (what scientists call the 'connectome'). Your brain would not be the same brain as it was when it was still attached to your body."[18] In fact, critics predict that head transplant recipients will experience mind-body dissonance of such magnitude that "insanity and death" are possible. It goes without saying that a body requires its brain in order to operate, but the brain, similarly, needs more than merely to have oxygenated blood pumped to it; it needs its body. That the connection of a brain to a body to which it is not accustomed could lead to death, no matter how technically splendid the merger, is perhaps the greatest sign of the intimacy and importance of the mind-body connection.

THE BRAIN AS A PREDICTION MACHINE

Somewhere in our evolution from unicellular organisms, we largely (but not completely) abandoned the method of reflexive, programmatic response to the environment in favor of the ability to do calculations tailored to the specifics of the circumstances. We can engage in such customized responses because we have a brain with the power to predict the consequences of a situation and of our actions.

That our brain is continuously predicting the future is evidenced by our emotion of surprise.[19] We all have a set of knowledge and beliefs that is employed as our unconscious minds continuously analyze information about our current circumstances in order to plan for what will come next. Surprise is evoked when you encounter an event that doesn't match your brain's prediction. That signals to your unconscious mind that its schema may be faulty and need revision, and it interrupts your conscious mental processing and shifts your attention to the unexpected event because the unpredicted encounter could possibly pose a threat.

The "predicting the future" I'm talking about here is a different process than one might use to predict the swings of the stock market or who will be the next congressperson indicted for misuse of campaign funds. It's more like *There's a rustling in the bushes. The last time I heard a rustling in the bushes a bear stepped out and tried to devour me. Therefore I'd better run.* Or, *There's a mushroom in the dirt. The last time I ate a mushroom like that, I had the worst stomach-ache I've ever experienced. I'd better not eat that mushroom.*

Those more modest, immediate, and personal kinds of predictions—concerning what is about to happen in the immediate vicinity and in the very next moment—are key to survival and one of the last things we lose as we age. For example, as I write this, my ninety-eight-year-old mother has a deficit in her ability to reason. As a result, if we're going out, she's unable to foresee that it may get cold later that evening and doesn't ask to bring her jacket. But she still has the ability to respond in the immediate moment, so as soon as it does start to get cold, she will ask me to get her one. Or if I place her cup of coffee too close to the edge of the table, she'll become agitated and ask me to move it so it doesn't get knocked over.

As you go through life, your brain is constantly making those immediate predictions, preparing you to take action if necessary, and one of the key ingredients in those calculations is your core affect. For while your senses provide information about your circumstances, it is your core affect that provides data about the state of your body.

It is surprising, given how powerful the influence of core affect is, how often we are not consciously aware of it. If we're distracted, we may not even notice for a while that we're cold, or hungry, or coming down with the flu. The capacity to remedy that, to build an awareness of core affect, is key to taking control of our thoughts and feelings. We all instinctively act to do just that, to alter our mental state through our bodies. We soothe ourselves with comfort food or a glass of wine, we pump ourselves up before a game or a spin class by listening to music, we go for a run to create that contented, relaxed feeling we have after we finish. Once we understand the importance of core affect and learn how to become aware of it by checking our core affect "temperature," we can act consciously and proactively to regulate and transform it and to understand its effect on our feelings and behavior.

THE HIDDEN INFLUENCE OF CORE AFFECT

We live in technological societies which demand that we make complex decisions about all aspects of life—about our relationships, jobs, investments, elected officials, medical care, and many other social and financial situations that extend far beyond the immediate place and time. Core affect influences these predictions and decisions, but it evolved when we were living far more primitive lives. Evolution works slowly, so what worked for the past 500,000 years is not necessarily the best approach for the past 500 or for today. And so, the influence of core affect today is not always beneficial.

Consider the case of Kamal Abbasi, who, after five years in prison, finally went before the parole board. He'd been convicted of charges stemming from the purchase of chemical ingredients that could be used to make a powerful explosive. The chemicals had been ordered from what turned out to be a dummy website, set up as part of a sting operation. Abbasi, who was then nineteen, hadn't been planning a terrorist act. A person he'd considered a friend had ordered the material from Abbasi's home computer and lied to him about the purpose of the purchase. But the judge at Abbasi's short trial wasn't swayed by his story, and he was pronounced guilty. Now, after having been a model prisoner, he was requesting an early parole.

Prisoners who come up before parole boards have been convicted for all sorts of things from minor felonies to full-blown murders. The hearing officers have just two options: to grant the prisoner's request and set him free on the grounds of past and predicted future good behavior, or to deny it.

In his hearing, Kamal didn't revisit his explanation about having been duped; the conviction was water under the bridge. He focused instead on making the case that in prison he'd been a model citizen. While incarcerated, he had never once gotten in trouble. He had done volunteer work outside the prison. He had taken online college courses. And he'd become engaged to marry his girlfriend at the time of the conviction, who'd been his child-hood sweetheart.

Kamal had been looking forward to this hearing every day for five years, working diligently, pinning all his hope for the future on it. Now, it seemed, the promise of putting his stupidity behind him and building a decent life came down to this single eleven-minute hearing, just before lunch. When the decision came, Kamal was devastated: he'd been denied.

After the hearing Kamal was racked with regrets about what he could have said or done but didn't. How could he have swayed those who were judging him?

What Kamal didn't know was that his chances of being granted parole were far less contingent on his actions over the past five years than on a condition that was seemingly irrelevant—the time at which his case was heard. Because his was the last case of the morning session, the chances that he'd be granted parole were virtually nil.

That's shocking but true. Parole officers hear dozens of cases each day and in each instance hold not only a prisoner's future in their hands but also the futures of those that prisoner may affect if they set him free. To deny parole requires little explanation; to grant it is taxing. It requires justification: the hearing officer must consider and accept convincing evidence of a prisoner's rehabili-tation and feel certain that no societal harm will result from his being set free. A wrong decision could result in a murder or other violent crime. The hearing officers are energetic at the start of the day, and after each break, but become worn down by the steady parade of case after case they have to consider in between and as the day wears on. By the time of their morning coffee and cookie hiatus, just before lunch, and at the very end of each day, the offi-cers tend to be hungry and exhausted, and their negative bodily state exerts a profound influence on their decisions.

The consequence is disturbing: in a recent study that has quickly become a classic, scientists gathered statistics on 1,112 cases involving eight hearing officers, with an average of twenty-two and a half years of experience on the job.[20] They found that, on average, the officers granted a parole request 60 percent of the time when that case was the first of the day, or the first after the coffee break or lunch. But as the graph on the facing page depicts, the rate at which parole was granted dropped steadily for each successive case, until, for the last cases before the next break, parole was almost never granted.

Because core affect reflects our bodily state, as we grow more tired and hungry, our core affect grows more negative. That affects our decision making—we become more suspicious, critical, and pessimistic—and we usually don't realize it. When the hearing officers were debriefed about their decisions, they presented rational reasons for each one. Never did they recognize or acknowledge the influence of their core affect, the emotions it led them to feel, or the decisions it led them to make. Because there is so little understanding of the impact of core affect on the decisions made at these hearings—decisions that are literally life changing for those who go before the board—this unjust system will continue.

Other research has uncovered analogous effects in many different contexts. In a study of twenty-one thousand patient visits to two hundred different clinicians, for instance, researchers studied doctors' decision to prescribe antibiotics. Patients suffering from what is probably a viral ailment often request them, even though antibiotics have no effect on viruses. In such cases it is best for doctors to resist their patients' demands, but that takes mental energy. The researchers found that at the start of the doctors' days they prescribed antibiotics to about one in four patients who desired them but didn't need them. That fraction rose steadily through the day until, by the end of the day, it was one in three.[21] Doctors go through many intense years of training before being able to practice, but still, like the parole officers, their decisions were influenced not just by the facts but by their fatigue.

Having money on the line doesn't seem to eliminate the effect. For example, another study examined the quarterly earnings calls of major corporations. That's a conference call between the management of a public company, analysts, investors, and the media

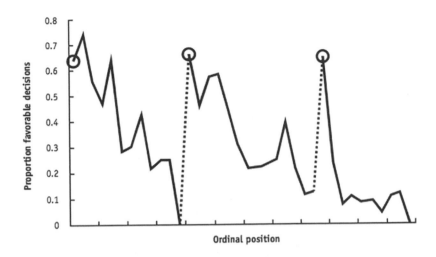

Proportion of rulings in favor of the prisoners by ordinal position. Circled points indicate the first decision in each of the three decision sessions; tick marks on *x* axis denote every third case; dotted line denotes food break.[22]

to discuss the company's financial results during the past quarter. The researchers found that, on average, the analysts and investors become increasingly negative as the trading day wears on, and that share prices decline in response to the negative effect of those late-in-the-day Q&A discussions.[23]

Whether we are hungry affects human behavior and decisions just as it did in the junco study. Abusive men in troubled relationships, for example, were found to be significantly more aggressive under conditions of negative core affect caused by low blood sugar.[24] Even foul-tasting food can have a negative effect. In one experiment, participants who were asked to drink a bad-tasting bitter liquid were judged more aggressive and hostile than a control group who didn't.[25]

For most of animal evolution, core affect has been a principal guide in our decision making, a key part of the apparatus that allowed organisms to survive the challenges of the wild by helping ensure that the body is tended to and functioning properly. We live in a safer world today, but core affect is still vital in guiding us to be aware of and tend to our bodily needs. It guides us toward resting when we are sleepy or ill, avoiding excess heat or cold, quenching our hunger and thirst. But as the above examples showed, a negative core affect can have unwanted side effects. You got a parking ticket in the morning, lost your credit card in the afternoon, and developed a headache in the evening as you were trying to forget the morning and afternoon. Due to all that, your core affect is anything but positive. Just then, your mother-in-law calls and suggests she come visit you next weekend. As you consider her request, your thoughts might overemphasize her tendency to comment on your weight, or the faded paint in your house, and undervalue the affection she brings.

A few years ago we had a fire in our house, and we were displaced for six months while the necessary repairs were being done. We lived in an apartment that was cramped, with beds that were not comfortable, and without access to most of our possessions, even the ones that were not lost in the fire. I tried to keep that in mind when my daughter, Olivia, made the usual teenager requests. I expected that my natural inclination would be much more negative than usual because the discomfort and disruption had influenced my core affect, leading me to say no to requests I would have granted in the past. Being a scientist, I tried to think of a way to test that suspicion. And then, the following April when I was doing my taxes, I realized there was one measure I could use—a quantitative one. I looked at my charitable giving for the half year I was out of the house and found that it was considerably less than usual. It's not that the fire was a financial hardship; our insurance company was very good about covering everything. But my core affect had been in a long slump.

My observation didn't constitute a controlled experiment, but it did get me thinking. Even when we are not going through a major life crisis such as a fire, death, or divorce, it is beneficial to keep in mind that our interactions and decisions—and those of the people with whom we are interacting—are all greatly influenced by core affect. What's more, neither we nor they are generally aware of its influence. The goal of mastering your core affect is best achieved by monitoring it, which will enable you to recognize how being cold or tired or hungry or hurting might be having an impact on you and how the same conditions might also be affecting those you interact with. Once you become aware, you can make a conscious effort to avoid situations analogous to that of the parole officers, in which you make bad decisions or have bad personal interactions that could have been avoided.

Our conscious experience is not formed from our brains alone; it depends also on how our bodies are doing and how we're treating them. In connecting our mental state to that of our bodies, core affect shapes our fundamental experience in the world and is thought to be one of the building blocks of our emotion. Here then, more than in the Platonic idea of rationality, lies the highest expression of our humanity. In the next chapter, moving back to the topic of emotion itself, we'll examine the interplay between emotion and rationality—the manner in which emotions guide our thought and reasoning.

PART II

Pleasure, Motivation,
Inspiration, Determination

How Emotions Guide Thought

Paul Dirac was one of the greatest physicists of the twentieth century, a pioneer in quantum theory who originated, among other things, the theory of antiparticles. As a quantum pioneer, Dirac played a key role in shaping our modern world, for the electronics, computer, communications, and internet technologies that dominate our societies are all based on his theories. Dirac's genius at matters of logic and rational thought made him one of the century's greatest thinkers, but equally striking was, as a young man, his utter lack of affinity or emotion in his interactions with his fellow human beings. Regarding people and their feelings, he proclaimed no interest. "I never knew love or affection when I was a child," Dirac told a friend, and he did not seek it as an adult. "My life is mainly concerned with facts, not feelings," he said.

Dirac was born in Bristol, England, in 1902.[1] His mother was British and his father Swiss, a schoolteacher famous for his harshness. Dirac, his siblings, and his mother were all bullied by his father, who insisted that the three children speak to him in his native French and never in English. Meals occurred in groups: Dirac's mother and siblings ate in the kitchen and spoke English. Dirac and his father ate in the dining room, speaking only French. Dirac had trouble with that language, and his father punished him for every mistake. He soon learned to talk as little as possible, reticence that continued into his early adulthood.

Dirac's academic intelligence, great as it was, provided little help in coping with the circumstances and challenges of everyday life. We humans evolved to exercise not purely rational thought but rational thought guided and inspired by emotion. Yet joy, hope, and love were largely missing from Dirac's coldly intellectual existence. Then, in September 1934, Dirac traveled to Princeton to visit the Institute for Advanced Study. The day after he arrived, he walked to a restaurant called the Baltimore Dairy Lunch. There, he saw his fellow physicist Eugene Wigner sitting at a table with a well-groomed woman who was smoking a cigarette. She was Wigner's divorced sister, Margit, a lively woman with two young children and no aptitude for science. She was known to her friends as Manci. As she would later recount, Dirac, thin and gaunt, looked lost, sad, disconcerted, and vulnerable. She felt bad for him and asked her brother to invite Dirac over to join them.

Manci was Dirac's antiparticle—talkative, emotional, artsy, and impulsive, while he was quiet, objective, and measured. Still, after that lunch, Dirac and Manci occasionally dined together. Eventually, "over ice cream sodas and lobster dinners," their friendship deepened, wrote Dirac's biographer Graham Farmelo. And then, a few months later, Manci returned to her native Budapest, while Dirac returned to London.

Back home, Manci wrote to Dirac every few days. Long letters full of news, gossip, and, most of all, feelings. Dirac answered with just a handful of sentences every few weeks. "I am afraid I cannot write such nice letters to you," he said, "perhaps because my feelings are so weak."

The lack of communication frustrated Manci, but Dirac didn't understand what was bothering her. Though the relationship remained platonic, they continued to write and occasionally see each other. With time, their attachment deepened. After returning from one visit with Manci in Budapest, Dirac wrote, "I felt very sad leaving you and still feel that I miss you very much. I do not understand why this should be, as I do not usually miss people when I leave them." Soon after that, in January 1937, they were married, and Dirac adopted Manci's two children. In his marriage Dirac achieved a level of happiness he had never thought possible. The Diracs remained the center of each other's lives until Paul Dirac died in 1984, shortly after the fiftieth anniversary of their meeting.

In a letter to Manci, Dirac wrote, "Manci, my darling, you are very dear to me. You have made a wonderful alteration in my life. You have made me human." Dirac's feelings for Manci awakened him. Being out of touch with his feelings, Dirac had been living half a life. After finding Manci—and his emotional self—he looked at the world differently, related to others differently, made different decisions in life. He became, according to his colleagues, a different person.[2]

Once he discovered emotion, Dirac grew to love the company of others, and—more important for our discussion here—he realized the beneficial effect of emotion on his professional thinking. In his mental life, this was Dirac's great epiphany. And so, as over the decades the most famous physicists of their generation approached the master for the secret to success in physics, what did he tell them? Farmelo ended his 438-page biography of Dirac on just that topic. Be guided, he reported Dirac advising, "above all, by your emotions."[3]

What did Dirac mean by that? Why would the cold logic of theoretical physics benefit from emotion? If someone were to take a poll of what people thought were the least emotional enterprises that humans engage in, theoretical physics would probably score high up on the list. But while it's certainly true that logic and precision are key to success in that field, emotion plays an equal role.

If skill at logical analysis were sufficient to achieve success at physics, physics departments would be staffed by computers rather than physicists. People think of physics as consisting of formulas like A plus B equals C. But as you're doing research, you usually run into situations in which A plus B could be C or D or E depending on which assumptions you choose or what approximations you make. And even the act of exploring what A plus B equals is a matter of choice. Maybe you should instead look into A

plus *C* or *A* plus *D*. Or maybe you should give up and look for an easier research project.

In chapter 2, I described how human thought at its most basic level is governed by fixed scripts, and how emotions arose as a more flexible way of reacting to novel circumstances. So too, in physics, does emotion guide your decisions about paths of mathematics to explore, based on both conscious and unconscious processes that encode your goals and past experiences in ways you might not realize. Just as the explorers of old used a combination of knowledge and intuition to find their way across the wilderness, physicists make their decisions based on the mathematics of their theory, but also their feelings. And just as the great explorers often pushed on with little to justify that decision, so too must physicists sometimes proceed with their grueling mathematical calculations fueled by little more than "irrational" exuberance.

If even the most precise and analytical thinking must be blended with emotion to be successful, it's no surprise that emotion has great influence over our daily thoughts and decisions. In life, rarely is there a clear and exact path or action to take. Our decisions are based on complex sets of circumstances and facts, probabilities and risks, and incomplete information. Our brains then process that data and calculate our mental and physical response. Like my father at the fence when he had to decide whether to join his fellow saboteurs, most of us, in making our decisions, are powerfully influenced by emotion and draw conclusions that may be hard to explain through mere logic. In what follows, we'll learn about the powerful role emotion plays in our mental processing—for both good (as with Dirac) and ill (as in the following story)—and what that means for us.

EMOTION AND THOUGHT

Twenty-year-old Jordan Cardella was devastated when his girlfriend broke up with him.[4] Some in that situation would have promised to change, or perhaps sent flowers. Cardella brainstormed a different way to win back her love. It was a method that one cannot help but think must have reflected the reason that she left him in the first place. His ex would come back to him, Cardella reasoned, if she found out that he was laid up, injured, in a hospital. It couldn't be some clumsy accident, though.

He needed to tug on her heartstrings. He decided to make her believe he'd been victimized. The key to Cardella's plan was sympathy, though the kind of feeling one bestows upon an abused dog is probably not the quality of romantic love Cardella wanted her to once again feel.

Cardella concocted a scheme. He asked an acquaintance, Michael Wezyk, to shoot him a few times in the back or chest with a rifle. In exchange he promised Wezyk money and drugs. He also asked his friend Anthony Woodall to call his ex-girlfriend afterward to tell her he'd been attacked by a group of men.

When the time came to execute the strategy, a few things went wrong. First, Wezyk refused to riddle Cardella with shots to the torso. Instead, he shot him once in the arm, then refused to fire again. Second, the police didn't buy their story. They charged Cardella with lying, and Wezyk and Woodall with being party to the negligent use of a firearm, which is a felony. Third, and perhaps worst of all, Cardella's ex didn't seem to care. She didn't show up at the hospital, or even inquire about his condition. Apparently, she didn't feel that the bullet hole in Cardella had remedied the shortcomings of their relationship.

The prosecutor said of the incident, "This has to be the most phenomenally stupid case that I have seen." Or, as the defense attorney Sanford Perliss put it, "felony stupid."[5] Cardella, afterward, probably agreed. That the plan didn't seem that ill-advised at the time he hatched it is a testament to the effect emotions have on our mental calculus. Cardella's intense love led him to a goal of winning back his ex at any cost, and it colored his thought processes to the point that when hatching his plan, he completely ignored common sense.

"An emotion is a functional state of the mind that puts your brain in a particular mode of operation that adjusts your goals, directs your attention, and modifies the weights you assign to various factors as you do mental calculations," says the neuroscientist Ralph Adolphs. Even when you believe you are exercising cold logical reason, you aren't, he tells me. People aren't usually aware of it, but the very framework of their thought process is highly influenced by what they're feeling at the time—sometimes subtly, sometimes not.

"Think of an iPhone," says Adolphs. In its normal mode of operation, the phone's goal is to always be ready to serve you. To best accomplish that, the iPhone is constantly at work, doing things like "listening" for you to yell, "Hey, Siri"; checking to see if there are new emails to download; and downloading new data to update your apps even if you are not currently using them. In low-power mode, the priorities are changed. Energy conservation is important, so these actions are reduced or stopped altogether. The phone still operates by executing logic-based calculations, but it is running a different program.

Though enormously more complex, the human brain is like an iPhone in that it is a physical system that carries out computations.[6] It evolved to compute what actions would be most likely to nurture your health, prevent premature death, and increase the probability that you will successfully reproduce. And like the iPhone, our brains possess a number of specialized programs, each tailored to solving a problem. Some of our programs apply to practical issues such as foraging, mate choice, facial recognition, sleep management, energy allocation, and physiological reactions. Others handle cognitive issues such as learning, memory, goal selection and priorities, behavioral decision rules, and probability assessments.

Just as the iPhone adjusts its programming when in low-power mode, our brain can run in various modes, each with different characteristics. An emotion is a mode of mental operation, a functional backdrop, that orchestrates and coordinates the brain's many programs in a manner tuned to the type of situation you are in and keeps the programs from conflicting.

I was once hiking with my son Nicolai in the rolling hills in the vast desert in Southern California. He was eight. It was getting late and I was feeling hungry and thinking about where to go for dinner. Then I realized that I had lost track of where we'd come from. Every direction looked the same to me, and I couldn't see far in any case, because my view was obscured by the neighboring hills. There was no one around, we were out of water, we were lost, and it would soon be dark and very cold. Suddenly I was afraid, and my hunger vanished. I wasn't ignoring my hunger; I no longer felt it. When you are in a state of fear, your senses are heightened, and distracting feelings, such as hunger, are suppressed.

I managed to calm myself enough to ponder what to do. Though I didn't consciously recognize any landmark or remember where I'd come from, I had a hunch about a particular direction, and so we marched off on that heading. It proved to be the correct choice. That's how your mind operates. Your senses give your brain input regarding the environment; your memory provides information about the past; your knowledge base and beliefs ground you with regard to how the world works. When you are faced with a challenge, threat, or other problem to solve, you employ all those resources to calculate your response. Some of this occurs within your conscious awareness, some outside it. But those mental calculations can proceed in many ways. Where do you focus your attention? How much weight do you assign to the various costs and benefits of a potential action? How much do you focus on the risks? How do you interpret ambiguous input and information? All of this mental processing is guided by your emotions.

Jordan Cardella's lovelorn mental state led him to a monumental miscalculation, but on average, and over aeons of time, our emotion states—whether love, fear, disgust, pride, or some other—have modulated our brain's answer to such questions in a manner that increased our ability to cope with the world we've lived in.

THE GUIDING ROLE OF EMOTION

Walking down a dark, deserted street at night, you think you see a movement in the distant shadows behind you. Could a mugger be stalking you? Your mind moves into a "fear mode" of processing. You suddenly hear with much greater clarity rustling or creaking sounds that you would not have ordinarily perceived or that wouldn't have registered. Your planning shifts to the present, and your goals and priorities change. That feeling of hunger you had vanishes; your headache is suppressed; that concert you'd been looking forward to later that night suddenly seems unimportant.

We saw in chapter 1 that an anxious state leads to a pessimistic cognitive bias; when an anxious brain processes ambiguous information, it tends to choose the more pessimistic from among the likely interpretations. Recall that fear is similar to anxiety but arises as a reaction to a concrete and present threat rather than the anticipation of a possible future danger. It is not a surprise, then, that fear exerts a similar effect on our mental calculations: as you interpret your sensory input, you assign higher than normal probabilities to alarming possibilities. Walking down that dark street, you wonder, is that the sound of footsteps behind me? Such questions now dominate your thinking.

In one illustrative study on fear, researchers induced fear in their subjects by exposing them to a grisly account of a fatal stabbing.[7] They then asked them to estimate the probability of various calamities, from other violent acts to natural disasters. Compared with subjects whose fear had not been stimulated, these subjects had an inflated sense of the probability of those misfortunes—not only related incidents, such as murder, but also the unrelated, such as tornadoes and floods. The grisly photos affected the subjects' mental calculus on a fundamental level, making them generally more wary of threats in the environment.

Now suppose, though, that you are a muscular individual well trained in self-defense. That person you thought you heard behind you springs out of the shadows and demands your wallet. You might now experience anger rather than fear. Evolutionary psychologists tell us that anger evolved "in the service of bargaining, to resolve conflicts of interest in favor of the angry individual."[8] When you are angry, your mental calculus inflates the importance you place on your own welfare and goals at the expense of those of others. In fact, there is an interesting (and enlightening) experiment you can perform on yourself, employing a tried-and-true method of dealing with anger. Next time you are angry, simply walk away. Disengage. Give the anger some time to dissipate. Then rethink the conflict. You'll see that you weigh the arguments differently now, with more understanding and tolerance of the other person's point of view.

Humans evolved in small social groups and had to continually engage in both cooperative and antagonistic interactions. In that context, an individual's anger creates incentives for others to appease him. In the case of our ancestors, the ever-present threat behind an episode of anger was aggression. Because stronger individuals had more to gain by fighting than weaker ones, and presented a more credible threat, one would expect that among our ancient ancestors stronger men angered more easily than the weak. And indeed, studies show that that is true even today. The correlation is much weaker for women, who are typically less inclined to fight.

Each emotion represents a different mode of thinking and creates corresponding adjustments to your judgments and reasoning. For example, imagine you experience an unexpected lack of warmth or affection from a person in whom you have a romantic interest. Is it really rejection, or is it due to a factor that has nothing to do with you, such as a temporary preoccupation on the part of the other person? How you think about such issues will be influenced in different ways by different emotional states. If you are in an emotional state of anxiety, when presented with an ambiguous situation of that sort, you will tend to select the more upsetting interpretation and perhaps start wondering what you've done wrong. Did you say something rude the last time you were together? Did you forget something you were supposed to do? Like all emotions, anxiety gone awry can cause problems, allowing your worry to overwhelm reason. The benefit of anxiety, on the other hand, is that sometimes the more negative interpretation is correct, and you would have missed it had you not been in an anxious state that made you reflect on what you might have done to cause the problem, and how you might be able to remedy the situation.

One of the most vivid illustrations of how emotional states can affect our mental calculations comes from the sad tale of a hunting trip near Bozeman, Montana, in the early 1990s.[9] Two men in their mid-twenties walked along an abandoned dirt logging trail through the midst of thick woods, talking about bears. They had set out that morning to hunt them but hadn't come across any.

The men were finally heading home. It was almost midnight, and there was no moon. They felt exhausted, nervous, and afraid. They still wanted to kill a bear, but this late, and in the dark, they now also feared such an encounter. And then, as the two men rounded a bend in the trail, they perceived a large object moving and making noise about seventy-five feet ahead of them. Both afraid and excited, they would have experienced spiking levels of adrenaline and the stress hormone cortisol in their bloodstreams.

The sights and sounds detected by our senses are not those that are perceived by our conscious mind. Instead, the sensory input goes to areas of the brain that receive raw information and is passed through several layers of processing and interpretation before we become aware of it. That processing and interpretation is influenced by our prior knowledge, beliefs, and expectations and by our emotion state. Had they not been in their scared and agitated state, and had they not had their minds focused on bear, they might have interpreted the noise and distant motion as having a benign origin. But on that fateful night, both men concluded that they had encountered a bear. They both lifted their rifles and fired.

The hunters' mental calculations, primed by fear to protect them from suspected danger, proved to be far off the mark. The "bear" turned out to be a yellow tent with a man and woman inside it. Fear of bears has no doubt saved the lives of countless humans who might otherwise have been attacked and killed, but not this time. The shaking of the tent and the sounds coming from it were caused by the couple making love. One of the bullets struck and killed the woman. The young man who fired that bullet was convicted of negligent homicide. Two years later he committed suicide.

The jury couldn't understand how a person could mistake a shaking tent for a bear, even after dark. But the jurors were not excited and afraid. We all interpret the world and our options within it through the calculations of our mind. Emotion evolved as an aid to tune those mental operations to the specific circumstances we find ourselves in. It's a system that developed over many millions of years. It works very well for the most part, but even when our ancestors were still living on the African savanna, it was surely not foolproof. And the flip side of the benefits conferred by emotion is the calamity that can occur when it goes awry.

THE SOCIAL EMOTIONS

No species is static, and as, over time, our ancient ancestors became more social, our emotional makeup evolved to accommodate and serve our more connected existence. New and more complex layers were added to our emotional repertoire, having to do with human interaction and social norms such as loyalty, honesty, and reciprocity.[10] These are the so-called social emotions such as guilt, shame, jealousy, indignation, gratitude, admiration, empathy, and pride.

Indignation, for example, often arises when one observes a person transgressing social norms. Gratitude and admiration arise when we experience someone fulfilling or exceeding them. Jealousy and shame seem to have arisen because, as human societies were evolving, an individual's ability to physically defend his or her interests was central to maintaining status and reproductive potential. If a male's mate was sexually unfaithful, and that became public, his peers would notice, and that would increase the probability that he'd be challenged in the reproductive or other domains. The male emotion systems of sexual jealousy and shame evolved as a response to compel individuals to resist such eventualities, while the female's strong need for the feeling of attachment was shaped by her role, emphasizing the importance of finding a committed mate who would help care for her offspring.

Jonathan Haidt, now a professor of ethical leadership at New York University, has made a career out of investigating the relationship between human moral reasoning and emotion. One of his most famous papers, cited more than seven thousand times in the academic literature, is titled "The Emotional Dog and Its Rational Tail." I've argued in this chapter that our seemingly rational thoughts, calculations, and decisions are inextricably entwined with our emotions, which act—usually behind the scenes—to alter our mental calculus. Haidt goes even further, contending that emotion, and in particular social emotion, is the primary driver of moral reasoning as well as other kinds of thought processes.

Much of Haidt's work centers on what role disgust plays in our lives. Researchers have found that the basic neural apparatus that governed disgust in the physical world was adapted to the social context. The emotion that had originally protected us from eating food that had gone bad expanded during our evolution to become a guardian of the social and moral order.[11] As a result, today we become disgusted not just by rotten food but by "rotten" people. Across many cultures, both the words and the facial expressions used to reject repulsive material are also used to reject socially inappropriate people and behaviors.

In one research article, Haidt describes how he and his colleagues arranged for volunteer college students to be paid in candy bars and asked them to evaluate the morality of various scenarios. The control group would do that in an ordinary laboratory setting, while the experimental subjects would do it in a work space "set up to look rather disgusting." Haidt had hypothesized that the subjects would mistake the physical disgust arising from their surroundings as social disgust related to the scenarios they read. If physical disgust can bleed into the social realm (and vice versa), this would be support for his notion that the two emotions are closely related.

In Haidt's experiment, the room for the control group was perfectly orderly and clean. The disgusting room contained a chair that had a torn and dirty cushion; a trash can overflowing with greasy pizza boxes and dirty facial tissues; and a desk that was sticky and stained, on top of which were a chewed-up pen and a transparent cup with the dried-up remains of a smoothie. If, when reading this, you're thinking, "sounds like a typical dorm room," then you know college students better than Haidt did. His article concedes that their attempt to disgust the students had failed. Those in the "disgust" group didn't find the room disgusting at all (based on a questionnaire that they were given).

Haidt and his colleagues had more success with another experiment that used a more reliable method to elicit disgust, even among students—fart spray. (It turns out you can order this online.) In that experiment, researchers released the spray in their experiment room shortly before a group of participants arrived and then handed them a questionnaire probing their attitudes regarding moral issues such as whether it is acceptable for first cousins to have sex or marry. They found that compared with other subjects who answered the questions in a room that didn't stink, these participants made harsher moral judgments.[12]

Despite the one misfire, Haidt's studies have generally been well replicated. In an experiment by another group, for example, the experience of disgust evoked by drinking a bitter liquid increased moral disapproval ratings of an ethical violation.[13] And conversely, thinking about moral transgressions led participants to perceive an unpalatable beverage as more disgusting than did a control group.[14] What's more, scientists have documented a correlation between a person's feeling of vulnerability to infectious disease and the person's negative reactions to people who appear unhealthy or are very old, or even to those who, like foreigners, simply look different.[15] They found the same tendencies in one particularly vulnerable group of individuals: pregnant women.

If Haidt is correct about social emotions being the basis of our sense of morality, then those emotions are crucial to our ability to cooperate and live together in societies. Just as the function of brain structures is often elucidated through the study of people with damage to them, the importance of social emotions in maintaining a well-functioning society is illustrated by what happens when a person lacks that type of feeling, as in psychopaths. In 2017, for example, a sixty-four-year-old former auditor, real estate businessman, and gambler named Stephen Paddock checked into adjoining suites on the thirty-second floor of the Mandalay Bay hotel in Las Vegas and, with the unwitting help of the bellhops, brought up five suitcases of weapons and ammunition. On the night of Sunday, October 1, he fired more than eleven hundred rounds into a crowd of concertgoers below, killing fifty-eight and wounding 851, including those injured in the ensuing panic. Police never found a motive, even after years of investigation. In fact, the episode seems to have been executed with the same ho-hum attitude that might characterize a trip to the grocery store.

About a year later, another gunman walked into a country music bar in Thousand Oaks, California, that was frequented by many college students, including some who had been at the Las Vegas concert.[16] He shot twelve people to death before killing himself. This gunman took the time to post on Instagram during the shooting, just as you might do during a set by a band you liked. "It's too bad I won't get to see all the illogical and pathetic reasons people will put in my mouth as to why I did it," he said in one of the posts. And then this gunman explained himself. In doing so, he likely illuminated Paddock's motive as well. "Fact is I had no reason to do it," he wrote. "I just thought $@#&, life is boring so why not?"

People speak of psychopaths as if they were "crazy," but the connotation of "crazy" is "irrational," and psychopaths are not that. The gunmen found it easy to kill because psychopaths lack social emotions such as empathy, guilt, remorse, and shame. As a result, their mental calculations are perfectly logical, but without emotional guidance, and so a psychopath who goes people hunting might have no more feelings for his victim than you would have for the clay target if you went skeet shooting.

The fifth edition of the *Diagnostic and Statistical Manual of Mental Disorders* lists psychopathy under "Antisocial Personality Disorders." The disorder seems to be associated with an abnormality of the amygdala, as well as parts of the prefrontal cortex, and is estimated to afflict between 0.02 and 3.3 percent of the population. If the prevalence were, say, 0.1 percent, in the United States, that would amount to a quarter of a million adults. Though random mass shootings have become more common, we're fortunate that the impulse for people hunting, even among psychopaths, is exceedingly rare. But a lack of social emotion does often lead psychopaths to disregard social norms and exhibit a pattern of antisocial, immoral, and destructive behavior. We'd all behave like that if not for our social emotions, and so evolution was wise to bestow them.[17]

DRIVES AS EMOTION

Darwin, and the vast majority of scientists in the century-plus that followed, focused on the emotions they deemed "basic" and were strangely closed to expanding their list. Studies of emotions such as frustration, awe, contentment, and even love were rare in comparison, and sexual arousal, thirst, hunger, and pain were classified as drives or motivational forces rather than emotions. But that has changed in recent years as many scientists have developed the viewpoint that emotions are "functional states." By that they mean emotions should be defined by the functions they serve rather than the anatomy or mechanisms that produce them.

Today most emotion scientists embrace a much wider array of emotions and acknowledge that even thirst, hunger, pain, and sexual desire, if not classic emotions, have much in common with them.[18] Hunger, for example, is an emotional mode that evolved to magnify the value we assign to procuring food, but it is actually more general than that. It amplifies the value that we assign to things in general: laboratory and field studies show that physical hunger increases the intention to obtain not only food but also nonfood objects.[19] We all know that we overbuy when we walk into a grocery store in a state of hunger, but we may not be aware that we also purchase more when we go hungry to Macy's.

In that regard, the effect of hunger is the opposite of disgust: studies show that while hunger drives acquisition, disgust encourages disposal, whether of food or other items. For example, in a study at Carnegie Mellon University, scientists showed volunteers either a neutral film clip or a clip from *Trainspotting* in which a character reaches into the bowl of an incredibly filthy toilet.[20] Afterward, the volunteers were given the opportunity to sell back a set of pens they'd been given at the start of the study. Those who had seen the neutral clip sold their pens back for, on average, $4.58. But those who'd seen the disgusting film clip were willing to part with them for far less, on average just $2.74. When quizzed about their decision afterward, the disgusted participants denied having been influenced by the *Trainspotting* clip and instead justified their actions with more rational reasons.

Sexual arousal is another "drive," now often viewed as an emotion, and in that context is studied for the effect it has on mental information processing.[21] For example, like fear, sexual arousal affects your sensitivity to sensory input that may indicate danger, but unlike fear sexual arousal doesn't raise your sensitivity; it lowers it. You may normally be alarmed by odd noises outside your door at night, but if they occur while you're having sex, you are far less likely to become consciously aware of them. Similarly, sexual arousal diminishes one's focus on goals unrelated to sex, such as having the cheesecake you were craving or avoiding pathogens.

One recent provocative study examined how sexual arousal causes changes in the male mental calculus by arranging for young male undergraduates at Berkeley to answer a series of questions, either in a non-aroused state or while sexually excited. Students were recruited through ads placed around campus offering to pay $10 per session for male students to masturbate in the service of science. Several dozen responded and were divided into control (non-aroused) and experimental (aroused) groups. Those in the control group merely answered the questions. Those in the arousal group were told to answer the questions while stimulating themselves at home to erotic photographs the researchers supplied.[22] The table below is taken from that study. Note how the subjects' judgments differed in the aroused versus non-aroused states. The answers below are the averages for each group, on a scale from 0 (strong "no") to 100 (strong "yes").

QUESTION	NON-AROUSED	AROUSED
Are women's shoes erotic?	42	65
Is a woman sexy when she's sweating?	56	72
Would it be fun to tie up your sexual partner?	47	75
Could you enjoy having sex with someone you hated?	53	77
Could you enjoy having sex with someone who was extremely fat?	13	24
Can you imagine having sex with a 60-year-old woman?	7	23

Related studies report that, as has been portrayed in many a film, the male's feeling of bonding with his sex partner, and his level of attraction, which rise rapidly with arousal, can plummet soon after climax.[23] Though these researchers examined only sex-related issues, other studies found that the thought processes of men in a state of arousal differ in other domains as well. For example, they tend to be less patient, and to place more value than usual on immediate rewards, such as money, over delayed gratification.[24]

What about the effect of arousal on women? From the point of view of evolution, one would expect women to react to arousal quite differently than men. A male animal's reproductive success is defined primarily by the number of fertile sex partners he has, and the sex act requires little investment. A female's reproductive success, however, requires that she invest much in the sex act and its results. Females must expend a large quantity of calories over the gestation period of their offspring. They also incur significant pregnancy-related health risks, and their opportunity cost is high. While a male can continue to impregnate other women, a female must forgo further reproduction not only while pregnant but, in mammals, usually also during the energy-intensive process of lactation, which can last for several years. As a result, females must be more choosy about their mates and less swayed by sexual arousal. As one scientist put it, "Lust and ejaculation can have profound effects on men's perceptions . . . it is adaptive for males to be able to be 'blinded by lust' . . . it would generally be maladaptive for women."[25]

Unfortunately, there have been fewer studies of the effect of arousal on women than on men. One study that was done on both men and women found, unsurprisingly, that both men and women who are aroused are more likely to opt for unprotected sex than those who make the decision when not aroused but that the effect of arousal is significantly greater in men.[26] Another study examined the effect of sexual excitement on a woman's sense of disgust. Of men, Freud wrote, "A man, who will kiss a pretty girl's mouth passionately, may perhaps be disgusted by the idea of using her toothbrush."[27] That same contradiction is even more powerful in women.

In women, the saliva, sweat, and body odors of others are among the strongest elicitors of disgust, and yet in sexual situations they can be attractive. Why? Researchers hypothesize that in order to encourage intercourse, arousal diminishes our disgust programs. To test that, they showed women either an erotic film made for women or a neutral film and then asked them to perform tasks such as taking a sip of juice from a cup with a large insect in it and picking dirty soiled paper out of a jar (unbeknownst to the subjects, the insect was actually plastic, and the feces on the toilet paper was fake). As expected, the sexually aroused women rated the tasks as significantly less disgusting than the women who had watched the neutral film.

One of the most important decisions a human being faces is the selection of a sexual partner, and our emotion of sexual arousal developed as a tool to be used in that process. Whether they outwardly exhibit it or not, both men and women experience a rapid physiological response to even a brief social interaction with an attractive member of the opposite sex. In both men and women, for example, contact with a pretty or handsome face causes cortisol and testosterone levels to shoot up.[28] But poor assessment and decision making carry great evolutionary costs for women. Accordingly, a woman's emotional system has been shaped by evolution to weigh mating decisions carefully and exhibit pronounced "choosiness" while still encouraging the individual to engage in the sex act.

THE PURPOSE OF JOY AND POSITIVE EMOTION

In August 1914, despite the outbreak of World War I, the Arctic explorer Ernest Shackleton and his crew departed from Britain on the ship *Endurance* for the Antarctic. He had a bold goal: to be the first to walk across the continent, traversing the South Pole to end up at the Ross Sea. But in January 1915, the *Endurance* became trapped in ice and drifted for ten months before the ship's timbers began to weaken and the water started pouring in, sinking the ship. Shackleton and his men got on three lifeboats and made camp on a nearby ice floe. The following April he and his crew made their way to nearby Elephant Island. There, they subsisted on seal meat, penguins, and their dogs. But Shackleton knew there was no chance they'd be rescued from the deserted island, so he and five of his men climbed onto one of the twenty-two-foot lifeboats and set off on an eight-hundred-mile journey across the rough, freezing sea to South Georgia Island. When they arrived two weeks later, gaunt and exhausted, they disembarked and prepared to trek across the island to a whaling station on the other side. No one had ever crossed South Georgia Island before, and it wasn't likely that they would make it either. As they set out, Shackleton wrote,

> we passed through the narrow mouth of the cove with the ugly rocks and waving kelp close on either side . . . as the sun broke through the mist and made the tossing waters sparkle around us. We were a curious-looking party that bright morning, but we were feeling happy. We even broke into song.[29]

Could this starving and frostbitten group about to embark on a potentially suicidal mission really have been happy? What role does happiness play in our lives?

The emotions I've talked about so far occur as a reaction to threatening events that demand response for the sake of survival or reproduction. If you're traveling with a big wad of cash, you're careful about flashing it in public. That decision to be cautious is encouraged by the mental state of fear. Your fear in that case is useful, because it reduces the chances that you'll be robbed. But if you just won a huge wad of money, and your mental state is happiness, what is the purpose of that feeling? How did their joy help Shackleton and his crew survive?

Research psychologists have only recently begun to examine the nature of "positive emotions," such as happiness. It's a category that cuts across the other two I've mentioned—social and basic emotions. In the psychological literature positive emotions include pride, love, awe, amusement, gratitude, inspiration, desire, triumph, compassion, attachment, enthusiasm, interest, contentment, pleasure, and relief.[30] Twenty years ago, all these were on the periphery of emotions research. Uncontrollable anger, chronic fear, and debilitating sadness were problems crying out for a remedy, but no one was complaining about suffering from too much awe or being paralyzed by joy. So although the evolutionary purpose of positive emotions was a mystery, there was little research on them. Then came a landmark 2005 paper by Barbara Fredrickson and Christine Branigan, then at the University of Michigan.[31] The paper described an experiment that lent credence to a theory Fredrickson had proposed, called the "broaden-and-build theory."[32] Positive emotion has been an area of intense study ever since.

The broaden-and-build theory offers an evolutionary reason for our positive emotions. When it comes to risk, the human brain must maintain a delicate balance. The brain is designed to help us to avoid danger and to focus on what might be hazardous in the immediate environment. This argues against risk and exploration. Most of our emotions weigh in on that side. They serve to protect us by inducing a mode of thought that narrows our perspective in order to promote quick and decisive action in potentially threatening situations. On the other hand, the brain is also designed to make us curious, to want to broaden our knowledge, to take some chances and explore our environment. Although that requires taking risks, that's how our ancestors discovered new sources of food and water, which came in handy when the old ones dried up.

States of positive emotion, Fredrickson observed, generally have the effect of encouraging a certain amount of risk. They are modes of thought that broaden our perspective and, she theorized, motivated our ancestors to take advantage of their *un*threatened moments—to explore, play, form social connections, take chances, and push into the unknown. That's what their joy at the beautiful Arctic morning did for Shackleton's group: it inspired them to push forward and trek on, eventually reaching the whaling station and then returning to save the comrades they'd left behind. That's what positive emotion is for, Fredrickson argued: it gave our ancestors a survival advantage because it kept them moving forward to new and better places.

Research shows that happy people are more creative, open to new information, and flexible and efficient in their thinking. Happiness, studies suggest, has the effect of encouraging you to push your limits and to be open to whatever comes your way. It also creates the urge to think outside the box, to explore and invent, and to be playful. In adults play encompasses intellectual or artistic activities, but juvenile play is mainly physical and social, helping to develop physical and social skills. Young African ground squirrels, for example, play by changing directions frequently as they run, sometimes by jumping straight up in the air, turning in mid-flight, and running off in a new direction after they land. These are maneuvers they'll be employing not just as juveniles but as adults, when they will be useful for emergency escapes, particularly from snakes.

Or take pride. It creates the urge to interact and share news of your achievement with others, and to strive for even greater achievement, which can enhance your future prospects. Interest, meanwhile, produces the urge to explore, to investigate in order to expand your knowledge base and your store of experience. That expanded capacity can then be drawn on to help meet future challenges, which in ancestral times meant finding water and food and escape routes or hiding places. In the modern world, it gives you the nimbleness to negotiate an increasingly risky, ever-changing environment, where yesterday's skills won't suffice to meet today's challenges.

Awe, in contrast, is an emotion that often arises in the context of religion or nature. It centers on two themes: the feeling of being in the presence of something greater than yourself, and the motivation to be good to others. It motivates you to broaden your focus from your own narrow self-interest to that of the larger group to which you belong, which has the benefit of enhancing your ability to become part of collaborative social groups and engage in collective action for the good of everyone. For example, in one study psychologists asked fifteen hundred individuals across the country to assess how much awe they felt on a regular basis.[33] In an ostensibly unrelated part of the experiment, they provided each participant with ten raffle tickets that gave them an opportunity to win a cash prize. They told the participants that they could keep all ten or, if desired, could gift one or more of the tickets to another individual who hadn't received any. Those who had reported experiencing more awe in their lives gave away 40 percent more tickets than those who were awe deprived. In another experiment, scientists took subjects to a "spectacular grove of Tasmanian blue gum eucalyptus trees" on the Berkeley campus, some more than two hundred feet high. They ushered others to an area outside a mundane-looking science building. In both cases, a confederate of the scientists walked nearby and stumbled, dropping a bunch of pens. Those who'd just been spending their time gazing at the spectacular trees helped the confederate in significantly greater numbers than those who'd been standing outside the building.

Whatever its other purposes, positive emotion is strongly correlated with good health and a longer life expectancy. A 2010 review of dozens of studies concluded that there are several pathways through which positive emotion exerts its beneficial effects—your hormonal, immune, and anti-inflammatory systems.[34] In one study health experts in London collected data on the well-being of hundreds of men and women between the ages of forty-five and sixty.[35] They assessed their subjects' positive emotion using a method designed by the Nobel Prize–winning psychologist Daniel Kahneman, author of *Thinking, Fast and Slow*. Kahneman realized that you don't get a very accurate picture by asking people if they are happy in life. Instead, you tend to get an answer that is reflective of how they feel at that moment, or of whatever event has just happened, or whether the sun is out. What they are reporting is a momentary feeling and not their general state. And so, Kahneman realized, it is better to ask specific questions at multiple moments and then analyze the data statistically, which is just what the researchers did. They arranged to call people at random times on their cell phones, several times a day, and each time ask how they were feeling at that moment. They found that the least happy, compared with the most happy, had about 50 percent higher levels of cortisol and other biochemicals associated with long-term risk factors for disease.

In another study scientists had more than three hundred volunteers participate in a similar emotions survey over a three-week period.[36] When it was over, they had them come into the lab, where the researchers gave them nose drops of a solution that contained a rhinovirus, the virus that causes the common cold. For the next five days the subjects lived under quarantine, each day seeing only one person, the scientist who came in and examined them for signs of a cold. The researchers found that the volunteers with the highest levels of positive emotion were almost three times less likely to develop a cold than those who had reported the least positive emotions. Happier people, it seems, are better equipped to fight illness.

The sum of all the research on positive emotion is that people who have plenty of positive emotion in their lives tend to be healthier and more creative and to get along well with others. Positive emotion makes us more resilient, strengthening the emotional resources needed for coping, and broadens our awareness, allowing us to see more options when faced with a problem.

Unfortunately, compared with our ancient ancestors, we have far less opportunity for physical activity, and play, and greatly diminished contact with nature, especially fields and forests.[37] Research has shown that these and other conditions of modern life, which scientists have dubbed "discord," serve to diminish our experience of positive emotion. The good news is that we are not doomed to live that lifestyle. It is the default for our era, but with effort we can counteract that by building habits that encourage us to feel more positive emotion.

For example, it helps to make a conscious effort to focus at least once or twice each day on aspects of our lives that are going well or that we are grateful for. It's also helpful to think of situations or activities that you enjoy—small, simple things such as listening to music, eating your favorite foods, or taking a hot bath—and making an effort to fit those activities into your everyday life. Being social also increases levels of positive emotion—nurturing relationships, interacting and communicating with friends, helping people, engaging in group leisure activities, and both giving and receiving advice and encouragement from others.[38] And then there is exercise, which not only promotes happiness but also lowers stress and provides many physical benefits. Positive emotions might have evolved to give our ancient ancestors a survival advantage, but to experience them still enhances our lives today.

SADNESS, THE ARCHITECT OF CHANGE

I've talked about the positive emotions, but what about sadness? That's not an emotion any of us welcome. What is its role?[39] People feel happy when they attain their goals, angry when they perceive an obstacle to goal attainment, and sad when they perceive a loss or an inability to maintain or attain a goal. Sadness seems to serve two key functions. First, a person with a sad expression on her face conveys a compelling message. Her downcast eyes and drooping eyelids, lowered lip corners, and slanting inner eyebrows have a prominent effect on observers. That communication of sadness to others signals that she needs assistance, and because we are a social species, that assistance often comes. We all know that when someone cries, it creates a soft spot in our hearts and we want to help them, even if they are adults.

The other function of sadness is to promote changes in thinking that help one adapt. As a mental state, sadness motivates us to do the difficult mental work of rethinking beliefs and reprioritizing goals. It broadens the scope of our information processing in order to help us understand the causes and consequences of our loss or failure and the obstacles to our success. And it is geared toward reassessing our strategies and accepting new conditions that might not be desirable but that we cannot alter.

The manner in which we process information when we are in a sad state helps us figure out why things are going badly and how to change course. Such thinking helps us shed unrealistic expectations and goals, leading to better outcomes. In one study that supports that conclusion, researchers simulated foreign exchange trading based on historical market data for a certain time period. Economics and finance students were given market information relevant to a certain point in the midst of that period and then asked to make trading decisions while being played music to steer them toward feeling either happy or sad.[40] Because this was a simulation and the researchers had data on how the foreign exchange market actually performed, they were able to judge the success of the student-traders. The sad participants made more accurate judgments and more realistic trading decisions than the happy ones and as a result profited more.

Of course, given the choice of feeling happy or sad, we'd all choose happy: though emotions are mental states that guide our thoughts, calculations, and decisions, they are also feelings we experience. In the science of emotion, the brain states corresponding to emotion are usually studied separately from those conscious experiences. In this chapter I've talked about emotion as a mode of mental information processing that colors our thought processes. In the chapter that follows, we'll study that other—conscious—aspect of emotion, our feelings.

5

Where Feelings Come From

My father slowed down in his later years. He'd pause every few steps and avoid all rigorous physical activity. This was due not to a lack of energy, or to the usual aches and pains of age, but, literally, to a matter of the heart. For that organ, so often associated with emotion, also carries the responsibility of pumping blood through and a pump is an energy-intensive device. So when the wall of my father's heart became the victim of poor circulation, his blood pump was impaired and he had to minimize activity in order not to overtax it.

Nature is a lenient master when it comes to our long-term well-being. She does not order us to forgo bacon and milkshakes or to exercise regularly. But in acute situations she may rule with a strong hand. Try to eat human feces, and you'll gag. Encounter a fierce animal, and you'll recoil. Walk briskly when your heart muscle is starved for blood, and nature will oppose you. In particular, when there's an increase in heart rate, the nerves in your heart muscle will send an intense alarm signal to your brain to cause a big jolt of excruciating pain. That pain is called angina pectoris.

In the mid-twentieth century, surgeons thought they had a clever new cure for angina. They theorized that if you tied off a certain artery in the chest cavity, blood flow through collateral vessels would increase, improving circulation to the afflicted area.

In physics you can often explore theories by scrawling some mathematics on a pad of paper. In medicine the patient is the paper. So doctors began to perform such surgeries. The theory seemed to be substantiated: patients reported significant pain relief. Soon surgeons everywhere adopted the procedure. Who needs a controlled scientific study to validate something that you already know works?

But there were a couple of clouds in the otherwise sunny surgical sky. Pathologists reported that when they performed autopsies on patients who'd had the surgery, they found no evidence of improved blood flow.[1] Though patients said the surgery worked, their hearts said it didn't. In addition, animal researchers who'd operated on dogs also reported seeing no effect. The improvement, doctors began to suspect, was all in their heads.

In 1959 and 1960 two teams of doctors investigated the apparent paradox in controlled experiments that, today, would be disallowed on ethical grounds: they performed both real and sham operations in order to compare the results.[2] In the sham operations, the doctors cut into a patient's chest to expose the artery in question but then stitched the patient back up without tying it off.

The two studies supported the notion that the surgeries worked for psychological, not medical, reasons. In one, three-quarters of the patients who received the real surgery reported relief from their angina pain. But so did all five of those who'd had the fake surgery. This is the placebo effect at work.

One of the fake surgery patients was quoted in the research publication as saying, "Practically immediately I felt better . . . I have taken about ten nitroglycerins [in the eight months] since surgery . . . I was taking five nitros a day before surgery." Another reported no angina pain, and said he was "optimistic" about his future, but sadly "dropped dead after moderate exertion" the following day.

The patients' degree of cardiac abnormality, the doctors noted in their paper, did not correlate with the degree of angina pain that they felt. Just as the degree of anger different people experience differs upon receiving the same insult, so too does the degree of pain people feel even when the magnitude of bodily harm is the same. And, just as some may take no offense at what others would grow furious over, some may experience no pain from damage that would torture others. The strong psychological component is why the placebo effect is so powerful in relieving pain.

The surgical treatment in question, called "internal mammary artery ligation," was eventually abandoned, and by the 1990s a less invasive and more sophisticated technique had been developed, called "stenting." A stent is a tiny wire cage that is threaded through the groin or wrist and inserted into a blocked artery in order to open it up and increase blood flow. Like the internal mammary ligation, patients reported good results from the stents, and the operation, costing between $10,000 and $40,000 in the United States, became commonplace, despite there being no large-scale controlled study to offer evidence of its benefits. Then, in 2017, *The Lancet*, a prestigious medical journal, published a paper explaining that, like the earlier ligation surgery, stenting worked no better than a sham placebo procedure.[3]

The ligation surgery had not actually relieved the physical source of the pain. The patients' damaged hearts sent the same distress signals both before and after, and yet the patients—whether the surgery had been real or not—tended to have a greatly reduced *experience* of pain. The stent procedure, too, seems only to have affected the patients' conscious perception of pain, and not the neural signaling that the perception was responding to.

"All cardiology guidelines should be revised," said one surgeon in response.[4] "It's impressive how negative [the result of the study] was," said another. "It's a very humbling study," said a third.

We still don't understand how placebos work, but we do know that the mechanism involves brain regions linked to emotional reactions. In the traditional view, emotions are considered prototypical responses to specific situations. If you are threatened, you feel fear; if you encounter something unexpected, you feel surprise; if you get a raise, you feel joy; and if you suffer bodily damage— a burn, a cut, or a potentially deadly lack of blood flow to your heart—your nerves will send a signal to cause you to feel pain.

Or so the theory goes. But humans don't work that way. If even a primitive feeling such as pain is not reliably correlated to the trigger that supposedly causes it, then what about other emotions?

THE DETERMINANTS OF AN EMOTIONAL STATE

The psychologists Michael Boiger and Batja Mesquita write of Laura and Ann, two women in a relationship:[5]

> Ann calls home to say she will be home late tonight because there is an official function at work. Laura would have liked to spend some time with Ann, and after spending several days looking after her when Ann was home sick, Laura feels some entitlement. She responds to Ann's phone call by saying that it is irresponsible to work overtime after having been sick, and that Ann should take it easy. Ann feels trapped: She is behind on work and convinced it would look bad to skip an official function just after having taken sick leave. To top it off, she feels ill-understood by Laura. Ann is so frustrated that she snaps at Laura for paternalizing and hangs up quickly. Laura in turn feels taken for granted, and underappreciated.

This snippet illustrates the complex and subtle interactions of emotion in daily life: both women responded emotionally to the situation in ways that reflected not just the events of the moment but those of the past few days (and perhaps other more complex aspects of their relationship, rooted in their history together).

That our emotional reactions are influenced by more than the immediate incident that triggers them is a hallmark of emotion. If you're in line at a store and a person cuts in front of you, a certain amount of annoyance is normal, but if you haven't eaten in a while, the resulting negative arousal can exaggerate your feeling and lead to a confrontation. Or if you're in a hurry because you're racing to a job interview and someone cuts you off in traffic, you might react with intense anger. You might attribute to that person selfishness and disrespect, whereas in a less aroused state you would have remained calm and assumed the person was merely being careless, or was perhaps late for an important appointment.

Emotions offer us the flexibility to respond to similar events differently, depending on past experiences, expectations, knowledge, desires, and beliefs. In the Ann and Laura incident, if the two women hadn't been under so much stress—Ann feeling behind on work, Laura feeling taken for granted and unhappy that Ann didn't seem to prioritize spending time with her—each of them might have reacted differently to the situation, with less hurt, less anger, less resentment.

The emotions we feel as a result of a circumstance or incident are the result of a complex calculation that takes into account the obvious implications of what has just occurred but also more subtle factors such as context and core affect (state of the body). One of the most enlightening illustrations of how emotions arise comes from a widely quoted paper by Stanley Schachter and Jerome Singer, titled "Cognitive, Social, and Physiological Determinants of Emotional State." In that publication, Schachter and Singer describe injecting their subjects with either adrenaline or a placebo, but telling them the shot was a vitamin called "Suproxin" that would affect their visual skills—the ostensible topic of their research.

Adrenaline causes increased heart rate and blood pressure, a feeling of flushing, and accelerated breathing—all symptoms of emotional arousal. One group of subjects was told that their feeling of arousal was a "side effect" of the Suproxin. Another group was told nothing. They experienced the same physiological changes as the first group, but had no explanation for them. A third group, the control group, had been injected with an inert saline solution and felt no physiological effect.

All subjects were then exposed to a situation that was designed to provide a social context. Each of them was asked to wait in a room in which there was a stranger who was ostensibly another experimental subject but was actually a confederate of the scientists. For half the subjects, this confederate behaved in a euphoric manner, apparently happy to be part of this important research; for the other half, the confederate acted disgruntled and complained about the experiment.

In the absence of arousal, the feelings expressed by the actors had no effect on the subjects; that is, those in the saline injection control group reported feeling no particular emotion. The Suproxin injection subjects who had been warned of the "side effects," and thus had an explanation for their physiological arousal, also reported feeling no emotion. But the Suproxin subjects who had not been warned about the side effect reported feeling either happiness or anger, depending on the behavior of the stranger they encountered. Their minds had apparently constructed an emotional feeling from the fact of their arousal and the context in which they'd felt it.

By putting their subjects in a simple and controlled laboratory setting, Schachter and Singer's experiment was able to illuminate the origin of emotion in a manner that is difficult to achieve in more complex real-world studies. In real life, we don't have to cope with random adrenaline injections. But physiological arousal can be incited by many everyday phenomena, and some of these have been studied in experiments that mimic the Schachter-Singer study, only employing means other than an adrenaline injection to cause the arousal.[6] Those experiments have demonstrated that exercise, loud noises, crowds, and being startled can all cause physical arousal that lingers for many minutes after the inciting event has ceased and, just like an adrenaline injection, can lead you to feel anger, joy, or other emotions associated with arousal, depending on the context. In other studies, scientists documented that both exercise and loud noises significantly amplify aggressive reactions in response to provocation. And, in another, researchers found that post-exercise arousal increased romantic attraction for an alluring member of the opposite sex.

CONSTRUCTING REALITY

The Schachter-Singer experiment is the converse of the placebo example: the placebo studies show that you *might not* feel an emotion (pain) even though you are in a state that commonly causes it (exertion in an angina patient), while the Schachter-Singer experiment shows that you *might* feel an emotion even though you are *not* in a situation that merits it, or merits feeling it as intensely. Such "misattributions"—feelings that are not appropriate to the circumstances you are in—are the emotional equivalent of optical illusions.

That the perception of emotion leads to a phenomenon parallel to one we encounter in visual perception is no accident: the manner in which your brain assesses a situation in order to make emotional sense out of it is analogous to the way it decodes your visual world; indeed it is representative in general of the way the brain operates. With regard to both the physical and the social world, one of the main lessons of neuroscience is that our perception of reality is something we actively construct, not a passive documentation of objective events.

There is good reason for that. Our brains have to take shortcuts because our conscious mental capacity is too limited to process the massive amount of information that would be required in order to perceive the world directly. Take your visual world. A single "snapshot" view of your environment—say, a digital photo—requires at least several million bytes of data. The information bandwidth your conscious mind can handle, on the other hand, has been estimated to be under ten bytes per second. So if your conscious mind had to understand the visual world based on the literal interpretation of millions of bytes of data, it would freeze like an overtaxed computer. To avoid being overwhelmed, your brain works with far more limited data and uses tricks to fill in the gaps in a manner analogous to what a graphics program might do to sharpen an image—except in this case the "sharpening" is far more sophisticated. In other words, what you see isn't a direct reproduction of what is there—your retina registers only a rather low-resolution image of the external world—but after your brain's unconscious processing, what you perceive is clear and sharp. And to accomplish that sharpening, your brain employs more than just the optical data; it also draws on those same factors that influence the construction of emotion—your subjective past experiences and your expectations, knowledge, desires, and beliefs.

In my book *Subliminal: How Your Unconscious Mind Rules Your Behavior*, I wrote about a classic experiment that illustrates that in the auditory realm. When you listen to someone speak, you hear just a selection of the full auditory data. Your unconscious sound processing centers then make guesses to fill in the gaps before making the perception available to your conscious mind. To demonstrate this, experimenters recorded the sentence "The state governors met with their respective legislatures convening in the capital city," planning to play it back to a group of experimental subjects. But before playing the sentence, they erased the syllable *gis* and replaced it with a cough so the subjects would hear "le-*cough*-latures." The experimenters warned their subjects that there was a cough in the midst of the sentence and supplied them with printed text so they could circle the exact place at which the cough had masked a sound. If the human experience of hearing was a direct reproduction of the auditory data, it would have been easy to identify the obliterated syllable. But none of the subjects could do it. In fact, so strong was these subjects' "knowledge" of what the word "legislatures" is *supposed* to sound like that nineteen of the twenty subjects insisted that there was *no* missing sound.[7] The cough didn't affect the subjects' conscious perception of the sentence, because their perception was based on both the actual speech and other factors, which their brains employed to fill in the missing sounds.

That perception is a construction is not true just of one's perception of sensory input, such as visual and auditory information. It is true of your social perceptions as well—your perceptions of the people you meet, the food you eat, and even of the products you buy. For example, in a study of wine, when wines were tasted blind, there was little or no correlation between the ratings of a wine's taste and its cost, but there *was* a significant correlation when the wines were labeled by price.[8] That wasn't because the subjects consciously believed that the higher-priced wines should be the better ones and thus revised whatever opinion they had accordingly. Or rather, it wasn't true *just* at the conscious level. We know because as the subjects were tasting the wine, the researchers were imaging their brain activity, and the imaging showed that drinking what they believed was an expensive glass of wine really did activate their pleasurable taste centers more than drinking a glass of the same wine that had been labeled as cheaper. That's analogous to the placebo effect. Like pain, taste is not just the product of sensory signals; it depends also on psychological factors: you don't just taste the wine; you taste its price.

In the case of feelings, the direct data feeding your construction of your emotional experience are circumstances, environment, and your mental as well as your bodily state—your core affect. These inputs are all integrated and interpreted employing the same sorts of tricks and shortcuts your brain employs in perceiving pain, taste, sounds, and other sensations—until, finally, you feel something. All that processing is a good thing, because the indefiniteness of the connection between triggering event and emotional response gives us an opportunity to intervene and consciously influence the emotions we feel—a topic we will explore in chapter 9.

THE CONSTRUCTION OF FEELINGS

There is a school of psychologists and neuroscientists today—the constructionists—that goes even further in their assessment of the looseness of the correspondence between triggering event and the emotion you experience. They call into question the validity of the very idea of discrete categories of emotion—fear, anxiety, happiness, pride, and so on.

One widely accepted point the constructionists make is that the emotion terms we use in everyday parlance do not really refer

to single emotions, but are more like catchall terms for whole categories of feelings. It's an observation that goes back at least to
William James, who, in his 1894 article "The Physical Basis of
Emotion," argued that there is an essentially infinite number of
distinct emotions, one corresponding to each possible state of the
body.[9] "Fear of getting wet is not the same fear as fear of a bear,"
he wrote. Today scientists can document such distinctions and
trace the precise brain activity associated with different variants.
For example, one dramatic experiment illustrated the fact that fear
of external threats such as snakes and scorpions and the fear of
internal threats, such as suffocating, though both called fear, are
actually distinct mental states, and even involve different patterns
within the brain.

The experiment was a study on patients with amygdala damage. The amygdala plays an important role in many emotions,
including fear, but not all types of fear. In this experiment, subjects who feel nothing when snakes and scorpions crawl up their
arms *did* experience fear and panic when breathing air infused
with a high level of carbon dioxide, which simulates the feeling
of suffocation.[10] As Lisa Feldman Barrett, one of the leaders of
the constructionist school, put it, "People group very different
instances [of emotion] into the same category and give them the
same name."[11]

Conversely, the constructionists point out that just as we
may fail to see distinctions between emotion states and lump
them together under the same name, we may also make distinctions where none exist; that is, the emotion categories we use can
sometimes overlap. For example, as I've said, fear and anxiety are
viewed as separate from each other: fear is seen as a reaction to a
specific thing or circumstance, while anxiety is seen as an unfo-
cused, future-oriented fear. But real-life situations can blur those
lines, so that fear can be difficult to distinguish from anxiety. If
you are very sick and worry about dying, some would classify that
as fear, others as anxiety, but whatever the language being used,
the emotion is the same.

Constructionists hold that the language we use to denote fear, anxiety, and all the other emotions we can think of, though widely used, has little fundamental meaning. They believe that, as young children learning language, we learn to lump various emotional experiences together in the conventional manner determined by our particular language and culture. There is an analogy with color. Most languages/cultures assign to colors a discrete, limited number of names, such as red, orange, yellow, green, blue, indigo, and violet. But physics tells us that there are an infinite number of colors, a whole spectrum that goes from red at one end to violet at the other with a continuum in between. The constructionists see the terms we use to describe emotions as being just as arbitrary as those we apply to colors.

There is a lot of cross-cultural research showing that the languages of various cultures often don't agree on which "fundamental" colors they assign names to. They may even have a vastly different number of colors for which they have words. Some of the best support for the constructionist point of view comes from analogous studies of words for emotion. Due to widespread travel and global communications, there is so much intercultural exchange and influence that it can be hard to find a culture that hasn't been greatly influenced by other cultures, but they do exist. One is the Ilongot tribe in the Philippines, which thrives in an isolated forest enclave and has resisted all attempts at assimilation and modernization. The Ilongot identify an emotion they call *liget*, which is unique to them, and for good reason: it describes the experience of intense, euphoric aggression that accompanies a head-hunting expedition.

There are also less exotic examples. Consider sadness and anger. These are experienced as separate emotions in the West, but in Turkey (and Turkish) they are considered one emotion, called *kizginlik*.[12] More generally, anger-like emotions are more common in Western cultures, which emphasize the autonomy of individuals, than in Eastern cultures, which emphasize harmony and interdependence.[13] The Tahitian language, meanwhile, has no word translatable as sad. One scientist described a Tahitian man whose wife and children left him and moved to another island.[14] The man said he felt "without energy" and considered himself ill.

The English language has words for hundreds of emotions. Other languages have far fewer—for example, just seven in the Chewong language of the Malay Peninsula. As one emotion researcher put it, "Different languages recognize different emotions. They carve up the domain of emotion differently."[15] That's not to say that different people experience different emotions, but rather that the categories of emotion identified in various cultures are somewhat arbitrary.

That supports the idea that our feelings are not, as Darwin believed, innate hardwired responses to a set of archetypal stimuli. In an article on this issue that I co-authored with Barrett and Ralph Adolphs, Barrett argued that science, so far, has not identified truly objective criteria to reliably determine whether a person or an animal is in one emotional state or another.[16] Adolphs, like the majority of emotion researchers, recognizes her point, but wouldn't go that far in dismissing the usual categories. As to which school of thought is correct, the jury is still out.

EMOTIONAL INTELLIGENCE

In the fall of 2018, a Thai man named Nakharin Boonchai was driving on a road near Khao Yai National Park when two elephants started to cross.[17] Boonchai hit the rear animal, injuring two of its legs. The elephant turned and looked at the vehicle. It paused, then stepped over to the car and stomped on it, killing Boonchai instantly. Was this emotionally charged road rage or a reflexive reaction to being physically threatened? Despite studies on the emotional lives of elephants, no one knows whether, or to what extent, an elephant has conscious feelings. But we humans do.

Though it might seem that our feelings should be obvious to us, we've probably all at times discovered that we had been ignorant of what we were really feeling or why. Clarity about our unconscious emotion states, our conscious feelings, and the role of our more general life circumstances is the first step toward harnessing emotions in our service, or at least preventing them from working against us. The goal, for a happier and more successful existence, is to use that self-knowledge to increase your emotional intelligence.

As I write this, my mother, wheelchair-bound, lives in an assisted living home. Despite being nearly one hundred, she is in good physical health, but for several years her mind has been declining. She still recognizes me and the family and we can talk about my childhood, but if you ask her to add nine and three she cannot do it. Ask her to choose between two things to eat and she can't form an opinion (unless the choice involves chocolate). Ask her who the president is or what country she is in, and she doesn't know. And yet when I walk in to pick her up at the home, she might immediately say, "What's bothering you?" Or, "Something is on your mind." And she is always right. It is uncanny: our emotional intelligence is so ingrained in us that it seems to be one of the last things to go.

"Emotional intelligence" is a term that has become so much a part of the language that we assume it has always been with us, but it was actually coined in 1990 by two psychologists—Peter Salovey, at Yale, and John Mayer, at the University of New Hampshire, whom I mentioned in the introduction. As they wrote at the beginning of their first blockbuster article on the topic, they presented "a framework for *emotional intelligence*, a set of skills hypothesized to contribute to the accurate appraisal and expression of emotion in oneself and in others, the effective regulation of emotion in self and others, and the use of feelings to motivate, plan, and achieve in one's life."[18]

"Is 'emotional intelligence' a contradiction in terms?" they asked. It was a natural question because, as discussed earlier, in Western thought the traditional view of emotion is that it acts as a disruption of rational mental activity, not a contribution to it. Until recently people believed that whatever rational abilities it is that IQ measures represent true intelligence, and anything else is irrelevant. But Salovey and Mayer rightly saw that emotion and rationality could not be separated and that indeed the most successful people in society were often those with high emotional intelligence. They also realized that, conversely, many of the highest intellects in business and society who have low emotional intelligence run into difficulty.

Consider, for example, a 2008 experiment conducted by Adam Galinsky, of the Kellogg School of Management at Northwestern, and three colleagues. The scientists had MBA students engage in a mock negotiation concerning the sale of a gas station.[19] They arranged that the highest price the buyers were authorized to pay was less than the lowest price the sellers were authorized to accept. But price wasn't the only thing to be negotiated; both buyer and seller had other interests, which, if properly considered, could result in a deal that would satisfy both parties.

Before the negotiation, one-third of the subjects were given generic instructions, another third were instructed to imagine what the other side was *thinking*, and the final group was instructed to imagine what the other side was *feeling*. Those who focused on the other side's thoughts or feelings were significantly more likely to strike a deal than those who did not. Negotiation is just one niche in the ecosystem of the business world, but over the past decades researchers have found that businesspeople with the ability to understand the feelings of others excel at management, human resource issues, leadership, and many other aspects of their profession.

Though often lacking, emotional intelligence is important even in the sciences, because conducting good research is unfortunately just the first step toward success in that highly competitive profession. At a time when there has been an explosion in the amount of research being done, the ability to get colleagues to pay attention to and understand your work is often at least as important as one's raw scientific ability.

Those who are not attuned to others can also have difficulty making friends. For example, they may be oblivious to social cues and continue talking even when a conversation partner wants to end the conversation or to chime in and respond, and they may not react appropriately when the other person speaks of a charged emotional issue. Emotional intelligence is so important to our species that it appears in human babies by the age of two or earlier. Toddlers of that age, if they see a family member in distress, will react by trying to help them or by breaking into tears themselves.

Just as our emotion states influence our information processing, they also influence our communication. Emotion is what lubricates our conversations and allows us to relate to each other and understand other people's wants and needs. Whenever we meet somebody, we send out emotional signals, and being able to read them involves both conscious and unconscious processes. People adept at emotional intelligence know how to monitor their own expressions of emotion and to attune themselves to the reactions of others. They are aware of the signals they send as well as those they receive, which allows them to communicate much more effectively. Those who are particularly adept at reading and connecting to others are the people we tend to think of as charismatic. Good leaders can communicate not just to a handful of people in their immediate midst but to large audiences, in person or even on television.

Not only are humans blessed with the ability to read other humans; we also want others to know us.[20] Studies reveal that 30 to 40 percent of people's conversation concerns their private experiences and personal relationships. On social media, 80 percent of posts are about people's immediate experiences. In fact, in a 2012 study at Harvard, researchers had participants talk to them, either about themselves or about others, while their brains were being imaged in an fMRI machine. The scientists found that engaging in self-disclosure activated brain regions associated with reward and pleasure significantly more than talking about others.

In another experiment, subjects were presented with a series of 195 questions and were told that they'd get a few cents per question answered. Each question came from one of three categories—a question about themselves, a question about someone else, or a question of fact—and before each question the subjects were allowed to pick the category. When the payoffs were equal in all three categories, subjects chose the self-category about two-thirds of the time. And when the payoffs varied among the categories, they continued to choose self-oriented questions more frequently even when it resulted in lower earnings than always choosing the category with the greatest payoff. Individuals, the scientists wrote, were "willing to forgo money" for the opportunity to reveal things about themselves.

Ours is a social species. We exist not alone but as part of a society. When a flock of birds changes direction, there is no driver bird telling the others what to do; they are coordinated through an innate connection of their minds, with each attuned to the others. That's true for us too. We are all connected, and those connections are made through our emotions.

During my brief time in the corporate world, I had two bosses, both executive vice presidents. My first was a woman who genuinely cared about those in her division and was good at reading their emotions and responding in an empathetic and constructive way. Her employees responded with loyalty and a willingness to go the extra mile whenever needed. When she retired, my boss was succeeded by a woman who was oblivious to how others felt. In one meeting she gave a pep talk in which she told us that her goal was for our group to become so profitable that within five years her annual bonus would break the million-dollar mark. No one was willing to go the extra mile to achieve *that* goal, and morale and profit plummeted. In the psychology literature, a person who understands what another person is thinking or feeling is called a "perspective taker." Those with the ability to take another person's perspective can smoothly navigate our collective emotional flight path, finding the right balance between competition and cooperation. Those who can't have a much harder time. So perspective taking is an important social skill, a key to charisma, persuasive power, and success in many areas, both professional and personal.

6

Motivation: Wanting Versus Liking

Within a period of about a year, Clara Bates, a young mother in Derby, England, was evicted from two properties due to the activities of her toddler daughter, Farrah.[1] What bothered Bates's landlords was understandable: Farrah was eating the carpets and walls. Bates first noticed the problem when Farrah was potty training and odd bits turned up you-know-where. Then Bates realized that those odd holes she'd noticed at the carpet edges were not due to normal wear and tear. Neither was the missing Velcro on her daughter's shoes.

Farrah's behavior is not without precedent. All these odd eaters suffer from an illness called pica, first described in a medical book in 1563.[2] The name comes from the Latin word for magpie, an intelligent bird in the crow family that is known to eat just about anything—seeds, fruits, nuts, berries, spiders, insects, bird eggs, baby birds, rodents, young rabbits, pet food left outdoors, random stray garbage. This is normal behavior for magpies. It is not, obviously, for humans.

The objects of attraction in pica sufferers often have to do with food, only at the wrong end of the food prep cycle; people crave not what is served on the plate but what is employed to clean it off. One man habitually drank liquid detergent. A woman executive had cravings for kitchen sponges. But it was Michel Lotito, a French entertainer, who was the Michael Jordan of the pica world.[3]

Lotito wasn't a fan of restaurants, but he enjoyed hardware stores. That's because he, literally, enjoyed hardware. Lotito hungered for metal. If it wasn't bite-sized, he would break it into small pieces and gulp it down with mineral oil and lots of water. Over time he is said to have eaten bicycles, shopping carts, and a Cessna 150 airplane, an expensive meal that he spread out over years. After four decades of such behavior, Lotito died in 2007, from an undisclosed "natural" cause. Presumably not an iron deficiency.

Why do we do the things we do? Why do we eat pasta and not pillows? Why do we eat at all? What processes within our brain cause us to take the actions we do, and how can we take control of or modify them?

Motivation can be thought of as the willingness to put effort into achieving a goal. It is a driving force that initiates and directs our behavior. Some motivations are biological, such as the motivation for food, driven by the homeostatic emotion of hunger. There are also social-based motivations such as the motivation for social approval and the motivation to achieve. These are likewise closely tied to emotion. In fact, the deep link between emotion and motivation is evident in the terms themselves: they have the same Latin root, *movere*. But human (and animal) motivation springs not directly from the neural networks that produce emotion but rather from a distinct neural system called the "reward system."

The reward system provides a flexible mechanism that allows our minds to take into account a wide variety of factors as we make decisions regarding when we need to act and for sorting through the possibilities in order to choose the most appropriate action. While more primitive life-forms act according to the fixed rules and triggers embodied in their innate programming, it is this more flexible, nuanced mechanism, responsible for much of our impulse to act, that makes us—and most other vertebrates—more than mere biological robots.

The new scientific insight into motivation sheds light on the cause of motivational disorders such as addiction while also illuminating how we might manage our urges and those of others. That understanding was largely achieved in two great leaps, decades apart. The first occurred in the 1950s, when the drive theory was finally abandoned due to the discovery of our reward system and the vast influence of reward system structures on other parts of the human brain.

SEEKING THE SEAT OF PLEASURE

If you read academic neuroscience papers, you get accustomed to sentences like this one: "We generated transgenic mice in which orexin-containing neurons are ablated by orexinergic-specific expression of a truncated Machado-Joseph disease gene product (ataxin-3) with an expanded polyglutamine stretch."[4] The article was on treatments for the sleep disorder narcolepsy, characterized by overwhelming daytime drowsiness; it seemed to me that the article could also *cause* narcolepsy. Given that I have come to expect such arcane descriptions of experimental procedure in academic journals, you can imagine how my jaw dropped when, while reading a 1972 article in *The Journal of Nervous and Mental Disease*, I came across this description: "On the afternoon of this study, the patient was again permitted to use the transistorized self-stimulation unit for three hours . . . He was then introduced to the prostitute."[5] The article, described by one subsequent author as "at once academic and pornographic," was penned by Robert G. Heath, who wrote an estimated 420 scientific papers over his forty-year career.[6]

Born in 1915, Heath began his career as a clinician, board certified in psychoanalysis, neurology, and psychiatry. In 1948, he became the senior psychiatrist on a research initiative at Columbia University aimed at improving the lobotomy as a treatment for schizophrenia and depression. In a lobotomy, surgeons essentially unplug a patient's prefrontal cortex by severing most of the nerve fibers connecting it to the rest of the brain. Knowing what we do today, scientists realize that such an operation deprives the patient of much of his or her humanity.

The prefrontal cortex, as we now understand, is a complex

and magnificent structure. Receiving input from numerous other brain regions, it plays an important role in conscious, rational thought. It helps to organize and focus our thinking. It coordinates our actions and goals. It censors unhelpful ideas and helps choose among conflicting behavior options. It is also responsible for our long-term planning skills, for reining in our impulses, and for helping to regulate our emotions. One of its subregions, the orbitofrontal cortex, is thought to be involved in the *experience* of emotion.

That's an enormous job description. But back when Heath was trying to "improve" the lobotomy, scientists didn't know all that about the functions of the prefrontal cortex. In fact, they didn't think the prefrontal cortex had much of a function at all. But they *had* noticed that removing the frontal lobes of a chimpanzee made it calmer and more cooperative. Which was how the Portuguese neurologist António Egas Moniz, who had noted similar "changes in character and personality" among soldiers who had suffered from injuries to their frontal lobes, came to invent the lobotomy in 1935. He received a Nobel Prize for it in 1949.[7]

Heath, like Moniz, was an enthusiastic believer in the new field of "biological psychiatry." That was an enterprise driven by the idea that physical brain abnormalities, rather than psychological trauma, cause mental illness. But Heath didn't think the lobotomy was very effective. It made the patients placid, and hence easier to manage, but it seemed to reduce symptoms by generally blunting emotion rather than achieving a cure for the underlying disorder. Eventually, Heath became convinced that the source of mental illness lay deeper in the brain, in less accessible, subcortical tissue—structures that are situated beneath the folded-napkin-like cortex. These structures had been shown to be important for emotion in cats. Of course, extrapolating from cats to humans would also lead one to conclude that people hunt sparrows in the backyard and like to sleep under the bed as well as on top of it. But in his extrapolations, Heath was essentially correct.

Ideas in science are a dime a dozen and gain value only when supported by experimental investigation. Unfortunately for Heath, the brain regions that intrigued him were too far below the brain's delicate surface to be reached through traditional surgery. And so his quest to find evidence to back his theory would be difficult and decades long.

His first attempts were based on a procedure a few doctors had begun employing the prior decade, the 1930s. In this new type of psychosurgery, they inserted thin electrodes deep into their patients' brains to take readings and to electrically stimulate an area or destroy it, depending on the illness being treated. Heath began experimenting with that procedure on animals, but he couldn't conduct any tests on humans. That's not because the obvious risk to the patient deterred him but because his colleagues, skeptical of his ideas, wouldn't provide him with the needed funding and logistical support.

Then one day Heath was lounging at the beach in Atlantic City and struck up a conversation with a stranger, a random meeting that would change Heath's life. The stranger, on vacation, happened to be the dean of the medical school at Tulane University in New Orleans. They hadn't yet introduced themselves when the dean started speaking of his work. He was in the process of setting up a new psychiatry department. He mentioned a researcher at Columbia whose work he admired. A guy named Heath, the dean said.

Today the process of securing a position as a professor is like a cross between running for mayor and applying to work at the post office. The hiring of faculty back then was simpler. No bureaucracy, no committees, no interviews, no politicking. If the dean met you at the beach, bathing-suit-to-bathing-suit, he could offer you a job on the spot. And that is just what the dean did.

At the time, Tulane neurosurgeons performed procedures at Charity Hospital. One of Heath's future colleagues described the facility as "a big sprawling beautiful hospital, containing some of the sickest patients you will ever see." Heath didn't care about the hospital's physical appearance any more than a kid would care if the candy store wall were covered with Picassos. The attraction to him was the endless supply of out-of-touch, disturbed (and sometimes violent) patients willing to sign off on any procedure that might bring them relief. "Clinical material," Heath called them.

Heath moved to New Orleans in 1949. Described by a colleague as handsome and charismatic, he soon persuaded the hospital to budget $400,000 to set up a 150-bed psychiatric unit. It would be his scientific playground, but he had a lofty goal: to employ the deep brain stimulation techniques he'd been using on animals to learn how to relieve the symptoms of mental illness in humans while at the same time studying the biological basis of the diseases. He was especially interested in schizophrenia.

At the time, the prevailing wisdom—derived from drive theory—was that people are motivated principally by the desire to avoid unpleasant feelings, such as hunger and thirst.[8] But Heath believed that reward, or pleasure, is just as important a motivator as pain. That view might have come from Heath's clinical training: Freud had decades earlier argued that pleasure plays a central role in human motivation. The "pleasure principle" hadn't been generally accepted by those studying the physical operation of the brain, but it appealed to Heath. And he took it a step further: he postulated that the brain must contain some discrete structure or structures that produce that feeling—some sort of "pleasure center." Schizophrenia, Heath theorized, was caused by a malfunction of that pleasure center. "Schizophrenics have a predominance of painful emotions," Heath said. "They function in an almost continuous state of fear, fight, or flight, because they don't have the pleasure to neutralize it."

Heath reasoned that if he could cause pleasure by stimulating the brain, he might relieve the symptoms of schizophrenia. He aimed to develop a way of permanently implanting an electrode and providing his patients with a means of stimulating themselves as needed, just as a person might pop a couple of aspirin for a headache. According to his contemporaries, he was obsessed not just with curing schizophrenia but with making a "spectacular" breakthrough, and perhaps because of that he was sloppy about the design, execution, and interpretation of his experiments.

One thing you may not want to hear said about your doctor is "he's ahead of his time." In the late 1940s, Heath was. Scientists were ignorant about where in the brain pleasure is produced; few scientists even believed there *was* a pleasure center in the human brain; and no one but Heath was looking for it.[9] As a result, Heath had little guidance regarding which structures to target. He was on his own, poking around in people's brains with his lead electrodes, relying on trial and error.

Without the benefit of modern technology, the placement of electrodes was very approximate in those days, and when inserted inaccurately, the electrodes could cause serious brain damage. Severe infections were also common. And worse. Among Heath's first ten patients, two died. Others suffered from convulsions. One, as the current was turned on, began to scream, then got up from the gurney, ripped at his clothes, and yelled, "I'm gonna kill you!"

Heath's attitude toward the dangerous complications seemed to be that these patients were severely ill and thus had nothing to lose. Indeed they *were* volunteers, and many might have agreed with that. Still, judged by today's standards, that may seem just a step up from the Dark Ages on the ethics scale. A neuroscientist friend once remarked to me about the "outrageous" acceptance of human experimentation in Western culture prior to around 1980, which seems not that long ago. Then a kind of scientific "me too" reform movement caused a reconsideration of the kinds of risks it was acceptable to expose experimental subjects to, and the standards have been different ever since. As a result, some of what was considered acceptable before the 1980s could land you in jail today.

Heath stopped his electrode-based schizophrenia experiments in 1955, not due to the human toll, but because his theory of schizophrenia proved wrong and the treatment didn't work. But like a mechanic who opens a muffler joint after his transmission shop fails, for decades to come Heath continued his haphazard electrode experiments, now on patients with other maladies, such as narcolepsy, epilepsy, and chronic pain. He also continued to investigate the effects on motivation and emotion.

Though wrong about the specifics, Heath was correct in general that major psychiatric illnesses have a physical origin. Unfortunately, the causes of schizophrenia and similar disorders would remain elusive for another sixty years. Pinpointing them proved difficult because scientists can't distinguish whether a patient is schizophrenic or bipolar by looking at a deceased patient's brain, nor are there any apparent differences when a sample of brain tissue is examined under a microscope. It wasn't until 2015 that, fueled by recent advances in genetics, scientists began to uncover the true roots of such illnesses. Much more work needs to be done, but we now know that they arise in patients having fewer genes involved in signaling between neurons and more genes related to neuro-inflammatory cells, which leads to low-level but chronic brain inflammation. An excess of dopamine production, related to the reward system, also seems to play a role but in a more complex and subtle manner than the pleasure deficit that Heath had envisioned. Such discoveries may eventually lead to effective treatments.[10]

Heath was far from the mark in his ideas about schizophrenia being caused by a malfunction of the pleasure center. But he would eventually be shown to have been on the right track with regard to the role of the pleasure center in motivation. And his belief that pleasure emanates from activity in specific regions of the brain would soon be confirmed. Today those regions are said to be part of the reward system, the key to human motivation. Unfortunately for Heath, due both to the limitations of technology and to his own undisciplined approach to experimentation, he himself was not destined to discover the system that he so passionately sought. Instead, shortly after he halted his schizophrenia experiments, the reward system was discovered by two young scientists who stumbled upon it while practicing their electrode placement skills on rats in a lab at McGill University.

WHERE MOTIVATION COMES FROM

Ironically, the same inability to accurately pinpoint the placement of electrodes that had plagued Heath resulted in a lucky fluke that worked in James Olds and Peter Milner's favor.[11] In 1953, Olds was a new postdoc. He had no experience working with rodent brains, so Milner instructed him. To hone his technique, Olds decided to implant an electrode in a rodent, aiming for a region near the base of the brain that was a popular subject of study in those days. He didn't realize it, but he missed.

After the rat had recovered from the operation, Olds tested the effect of stimulating its brain. He placed the rodent in a large box and sent a small jolt of current through the electrode. He noticed that afterward the rat sniffed around the region of the box where the stimulation had occurred and would return there if moved away from it. He also noted that if he started to stimulate the rat's brain while the rat was in a different part of the box, the rat would race there instead. In fact, he found that he could motivate the rat to go anywhere in the box by providing it with brain stimulation while it was there. It seemed as if the rat enjoyed the stimulation and were returning for more.

Upon x-raying the rat's brain, the researchers discovered that Olds had inserted the electrode into a then-obscure structure deep within the brain, called the nucleus accumbens, or, more simply, the accumbens. The accumbens is an important limbic system structure located deep in the brain. There is one in each of the brain's hemispheres. In humans, each accumbens is about the size of a sugar cube or a marble.

Olds and Milner procured new rats and inserted electrodes into their accumbens to see if they could replicate the effect they'd found in the first rat. They did. Next, they arranged things so that the rats could stimulate the electrodes themselves, by pressing a lever. To the scientists' astonishment, the rats became obsessed with self-stimulation, pressing the lever dozens of times each minute. They lost interest in everything else—mating, even eating and drinking. If left with plentiful water and the lever, they would continue to press the lever until they died of thirst.

The researchers theorized that the rats became obsessed because the accumbens played a role in their feelings of emotional pleasure. It seemed that, just as Heath had believed, the rat brains had a pleasure center and that feelings of pleasure motivated the rats even more than their survival drive. The scientists began to investigate which other brain areas would inspire self-stimulation. They uncovered several, running along the brain's midline and connected by a massive bundle of nerve fibers, all parts of what we today call the reward system.

Olds and Milner concluded, like Heath, that the attainment of pleasure is our primary source of motivation. They published a paper on their research, calling it "Positive Reinforcement Produced by Electrical Stimulation of Septal Area and Other Regions of Rat Brain." Their local newspaper, *The Montreal Star*, had a more sensational take on their work, running the headline "McGill Opens Vast New Research Field with Brain 'Pleasure Area' Discovery: It May Prove Key to Human Behavior." It was the spectacular breakthrough that Heath had dreamed of, but it had been made by others.

That brings us back to the paper describing the man and the prostitute. Olds and Milner's discovery inspired many scientists to experiment on animals. Heath, with his own brand of ethics, was among those influenced by their work. Having abandoned his work on schizophrenia, he decided to follow up on Olds and Milner's work with experiments of his own. He began inserting electrodes into the accumbens and surrounding areas just as they had—but not in rat brains, in *human* brains.

Heath had finally developed a way to keep the electrodes in place even while a patient moved about, creating the possibility of studying patients in a real-life environment, and the real-life context that most excited Heath was sexual. In that 1972 paper, Heath described a number of experiments in which he combined electrode stimulation with pornographic films, and in one case, the services of a prostitute, so that he could monitor his subjects' brain waves during orgasm. Heath was successful in producing orgasms, but not in understanding their mechanism.

The scientific method exists for good reason: it restrains you from jumping to false conclusions and guides you toward valid ones. Science generally advances in tiny steps, not great leaps.

Unlike our theories about everyday life, in science each idea and hypothesis must be made precise, and each experiment conducted with meticulous precision. After an especially good performance, a basketball player may become convinced that he performs better when wearing a certain mystical pair of socks. But to convince a scientist, you'd have to quantify what you meant by "better," play many games with the mystical socks as well as with other socks, and analyze the resulting statistics. That's why Magic Johnson was a great nickname for a basketball player, but as a scientist I would not be flattered to be called Magic Mlodinow.

Careful science, however, was not Heath's style. Heath had some of the most important traits of a great scientist: he was smart and creative, a visionary with true insight into the physical processes that create motivation. But he was sloppy and reckless. To consider him a great scientist despite that would be like saying, "He's a great chef, although he burns all the food." Though a pioneer in the role and mechanism of pleasure within the brain, Heath was outside the mainstream in both his ideas and the methods he chose to investigate them. And so despite being a pioneer with promising theories, Heath produced more than four hundred published works that shed little light and are now mere scientific curiosities, while it eventually fell to others to deliver on the potential of his ideas.

THE JOY OF BEING A MAMMAL

Animals constantly face situations that provide both opportunities and challenges. They forage and hunt, and they are hunted. If they are to survive, they must process the cues they receive from their environment and internal physiological state to produce effective behavior. That's the purpose of our motivational systems.

The most primitive life-forms successfully reacted to their environment without the help of a neural motivational system, or even neurons. Bacteria, for example, don't have a reward system. They act not because they seek pleasure but because they encounter a molecule that triggers an automatic response. As discussed earlier, they sense and respond to the presence of nutrients and, when placed in a nutrient-scarce environment, band together and cooperate to improve their efficiency. They snub neighbors that consume energy resources but fail to contribute. They defend their territory against other groups of bacteria, and they adjust their "battle" strategy according to their relative numbers. They accomplish their feats of survival by emitting and absorbing a wide array of molecules. The success of that approach is reflected in their numbers. In the human body, for example, there are more bacterial cells than human cells. That is no anomaly: the biomass of bacteria on earth exceeds that of all plants and animals. So although you may think of humans as the kings of the food chain, humans could equally well be thought of as mobile bacteria farms.

As effective as bacteria are, with no reward system a bacteria colony can only respond to stimuli in an automatic manner, as if they were biochemical Rube Goldberg machines. That approach is inherently inflexible, hence limiting. Organisms like the planaria, which 560 million years ago were among the first to have nervous systems, finally graduated from a reliance on preprogrammed responses. These creatures had a new ability: they could assess novel situations and react with actions that were tailored to specific circumstances and goals.[12]

We see the rudiments of a reward system in even the simplest of those multicelled organisms. The nematode roundworm C. elegans, for instance, with only 302 neurons, integrates sensory input using a centralized nervous system and employs dopamine, one of the neurotransmitters characteristic of our reward system, to drive its food-seeking behavior.[13]

With the evolution of vertebrates—reptiles, amphibians, birds, and mammals—came the more complex reward system architecture that humans share today. The vertebrate reward system is an all-purpose motivational network, activated in similar ways by different kinds of pleasurable stimuli. It is more sophisticated in mammal brains than in those of other animals.

A bacterium sensing nutrients nearby is programmed to go

after them and to avoid useless or harmful molecules. A (healthy) human finds eating an orange more *satisfying* than chomping on a Cessna 150. A bacterium's biochemical makeup determines whether it absorbs a molecule that floats past. A mammal *decides*. That's the operational advantage of the reward system. We don't react automatically; we weigh a variety of factors before choosing our actions. Our brain evaluates the pleasure of each potential experience and weighs that against the possible costs, employing what it knows, through core affect, about our bodily state, the future consequences of various actions, and other relevant data. Only after that analysis does our brain decide on our goal and motivate us to act.

Robert Heath retired in 1980. By then, through decades of painstaking research, others had worked out many of the details of the reward systems in humans and many other animals. By the mid-1980s, psychology textbooks explained that the reward system is a joyful set of structures that motivates us through feelings of pleasure to take the actions necessary to survive and thrive. According to that theory, we avoid that which brings pain and discomfort, and take action to maximize enjoyment, ceasing when a satiety feedback loop within the brain diminishes the pleasure the reward system provides. That's why we're driven to eat, and eventually to stop eating, chocolate and cheesecake.

The reward system theory accounted for motivation far better than the old drive theory. But some researchers, especially those studying addiction, still ran into questions the model had trouble answering. For example, some addicts continued to take drugs even though they reported no longer liking the effect. What motivated them? No one knew. But no one questioned the reward system theory, either, until a lone scientist, saddled with an experiment he couldn't make work, eventually realized the failing was not in his laboratory method. It was in the theory his experimental procedures were based upon. And so began the next revolution in our understanding of animal reward systems. The degree of pleasure that we experience, this scientist realized, is only half the reward system story.

WANTING VERSUS LIKING

The new reward system revolution would redefine psychologists' understanding of the connection between our pleasures and our desires. It has always been clear that we might decide not to seek something we like because we know it is unhealthy or we judge it to be unethical. That's an example of using our conscious will to control behavior. But it doesn't mean that, health or ethics aside, we don't like the thing we're depriving ourselves of. To pass up the brownie so you can fit into your slacks doesn't signify a lack of lust for chocolate, only an ability to rise above the lust, and psychologists had always believed that our capacity to defer or decline enjoyable experiences doesn't change the fact that we want to experience them. That we want what we like, and that we like what we want, seemed axiomatic. The fact that it is indeed not true required almost thirty years to gain acceptance.

The first step in this understanding came shortly before Christmas 1986, when Kent Berridge, then a young assistant professor at the University of Michigan, received a call from Roy Wise. Over the preceding decade, Wise had done groundbreaking work on the role of dopamine in the reward system, producing the kinds of studies that landed dopamine in the news with the label "pleasure molecule."[14] He wanted to team up with Berridge because Berridge was an expert at interpreting rats' facial expressions. By looking closely at a rat, Berridge could detect emotions ranging from joy to disgust. It was an odd kind of expertise, but Wise had in mind an experiment on pleasure, and there weren't many people around who could determine when a rat was having a good time (nor many who'd wish to). But Berridge wrote a recipe book on the subject, a twenty-five-page literature review that attracted more than five hundred citations in academic journals.[15]

A rat's brain, though having the same basic structure, is far simpler than a human's, and so is a rat's psychology. To a rat, any cage set up to dispense sugar water is a three-star Michelin restaurant. Wise reasoned that if dopamine is truly the pleasure molecule, impeding its action ought to make the sugar water no more pleasurable than wet sawdust. So he planned to inject rats with a drug that blocked that neurotransmitter and to compare rats' reactions to the treat before and after the dopamine blocker was administered.

Wise expected that before the blocker the rats would poke their tiny tongues out and lick their lips with pleasure, as is their habit in such situations. After the blocker was injected, Wise hypothesized, the pleasure reaction would be diminished. But how to quantify the change? That's where Berridge's expertise would be important: the frequency of licking is an indication of a rat's degree of pleasure and can be measured by a specialized instrument called a "lickometer." Berridge, who marveled "at the beauty" of Wise's work, was excited to team up with the famous scientist.

The experiment failed. The rats made the same pleasure faces both before and after the dopamine was blocked. If this were a Hollywood movie, Berridge would go home that night despondent, stare into the fireplace, and have a dramatic epiphany that explained everything. In reality, the scientists didn't take their failure too seriously. "Sometimes you do an experiment, and it just doesn't work," said Berridge. You try again. He did. But there was still no difference in the rats' reactions.

Wise eventually lost interest. But Berridge, younger and perhaps more open to new ideas, tried once more, this time employing a powerful neurotoxin that attacks dopamine and "completely takes it out." The rats' joyous tongue poking and licking did not cease. But now Berridge noticed something else that was odd. Though the dopamine-blocked rats seemed to still enjoy the sugary treat, they would not take action to drink it on their own accord. In fact, if not force-fed, the dopamine-blocked rats would starve to death. Their *enjoyment* of the sugar water was not obliterated, but their *motivation* to drink it was.

Berridge's experiments seemed to contradict the accepted wisdom that it is pleasure that drives us. They also seemed to violate common sense. How could it be that food brought the animals pleasure, yet they did not seek any?

Berridge theorized that in our reward system there is a distinction between liking something and the motivation to seek it, which he called "wanting." We tend to want what we enjoy, but, he asked, is that connection a logical necessity? Can you enjoy something but not have any motivation to obtain it?

Think of programming a robot. In the robot brain the degree of pleasure "felt" in any given situation might be represented by a number in some register. The program would provide a recipe for what brings the robot pleasure and quantify how much pleasure each pleasure trigger brings, and for how long. The robot's degree of pleasure—the number in its pleasure register—would vary with time, depending on the robot's experiences.

Say the robot is walking outside when it accidently encounters something its programming defines as pleasure such as the faint fragrance of distant roses. To move toward the roses might make the fragrance more distinct and increase pleasure, but to initiate a new action requires a decision or command. So unless the robot's programming also includes an instruction such as "take action to increase your pleasure level," the robot won't change course to approach the rose. That would require two systems—one to define "pleasure," and another to control "wanting," the circumstances that trigger action to obtain that which increases the pleasure register.

That's what Berridge's experiments led him to realize about his rats: liking—that is, pleasure—and wanting/desiring—that is, motivation—are produced by two distinct but interconnected subsystems within our reward system.[16] Berridge speculated that humans are built that way, too. We have a "pleasure register" in our reward system—our "liking" circuit—but we have to be programmed to pursue what we like. And so we have a separate "wanting" circuitry in our reward system to determine whether we are motivated enough to pursue any particular instance of pleasure.

At least a hundred neurotransmitters have been identified in the human brain. For the most part each neuron is specialized in that it employs just one neurotransmitter to send its signals. If the wanting system runs on dopamine but the liking system does not, Berridge reasoned, that could explain the results of his experiments: by blocking dopamine, he was knocking out the rats' wanting system but not the circuits used for liking. If he was right, dopamine would be not the "pleasure molecule" but the "desire molecule."

Berridge sought evidence for his hypothesis. He had produced creatures that liked their sugar water meals but didn't want them. Could he produce rats that wanted the meals but didn't like them? Yes: employing a tiny electrical current to stimulate the rats' wanting circuits, he induced them to guzzle down a bitter quinine solution that, judging by rats' facial expressions when they drank it, the rats found unpalatable.[17]

That was strong evidence that wanting and liking operated independently in the brain, but Berridge went even further. He found that the liking subsystem employs opioids and endocannabinoids—the natural brain versions of heroin and marijuana—as its neurotransmitters. That's why taking those drugs amplifies sensory pleasure: they are the real "pleasure molecules" of the brain.[18] And when Berridge blocked those neurotransmitters, his rats behaved as he hypothesized: they no longer appeared to like their sugar water meals, but because their dopamine-based circuits were intact, they still wanted them.[19]

Berridge went on to look for evidence for such dissociations between wanting and liking in human behavior. In hindsight, they are easy to find. One example occurs in people who are addicted to drugs such as nicotine and desperately want their fix, even when it produces little or no feeling of pleasure. A more innocuous example is when attractively displayed items in a store boost your desire to possess them, even though you don't "like" those items any more than before you saw the display. In fact, the focus of advertising is to stimulate not your enjoyment of an item but your desire to have it.[20] Sometimes accomplishing that is as easy as simply putting it, or just an appealing photo of it, in front of you. In one experiment participants' brains were imaged while they were shown attractive pictures of high-calorie food. The food images stimulated their "wanting" circuits—in some people more strongly than in others. In a follow-up, the subjects were enrolled in a nine-month weight-loss program, and those who'd exhibited the greatest response to the pictures had the most difficulty losing weight.[21] Scientists can use such data to help predict, through brain imaging, whether your diet will work.

A common cause of wanting/liking mismatches arises from the very struggles we go through to achieve what we want. Psychologists have found that when we run into barriers in a quest to attain something, we sometimes want it more while liking it less. In 2013 a group of scientists in Hong Kong tested that on sixty-one male college students who thought they were engaging in a speed-dating exercise.[22] The researchers wanted the students to feel that they had input regarding whom they'd be dating, yet because this was to be a controlled experiment, they wanted all the students to date the same woman. So a few days before the event, they sent the students profiles of four women and told them to choose one, but the profiles were designed to make one of the women seem significantly more appealing, and, as intended, all the students chose her. The dates were then arranged.

The woman whose profile each student favored was in reality a confederate of the researchers. She was told to be responsive to some of the volunteers, smiling a lot, seeking topics of mutual interest, and asking questions to signal her interest. Researchers dubbed that the "easy to get" condition. When interacting with the others, she was told to be more standoffish and occasionally decline to answer the volunteer's questions. They called that the "hard to get" condition.

After the encounter, the participants were asked to rate, on a 1 (very negative) to 7 (very positive) scale, how they felt about their speed-dating partner. They were also asked, on the same scale, to report "the strength of their motivation to see her again." Not surprisingly, the male students assigned to the "easy to get" condition liked the woman significantly more. But it was the students in the "hard to get" condition who were more interested in a second date. The young male students liked the easy-to-get woman, but they wanted the hard-to-get one. Some two dozen centuries later, that study has finally validated the advice of that famous dating consultant Socrates, who advised the courtesan Theodote that she would attract more friends if she would sometimes withhold her affections until men are "hungry" with desire.[23]

MAPPING WANTING AND LIKING IN THE BRAIN

Kent Berridge spent years mapping the anatomy of the "liking" system. He and his team plotted the sources of enjoyment by administering microinjections of opioids throughout the brain and noting which spots enhanced the rats' pleasure, as gauged by their tongue wagging.[24] He discovered that liking doesn't arise in one principal structure, but is dispersed in a collection of small chunks of tissue spread throughout the reward system. In humans, each chunk is about half an inch in diameter. Berridge dubbed them "hedonic hotspots."[25] Some are deep within the midbrain, in structures such as the accumbens and the ventral pallidum (a structure that was identified and named by anatomists only a decade or so ago). Others are in the orbitofrontal cortex, which produces the conscious experience of pleasure.

Berridge found that the accumbens is the key structure of our wanting system, which is far more centralized than our circuits for liking. Whenever we feel compelled to eat, drink, copulate, sing, watch television, or exercise, signals from the neurons in our marble-sized accumbens are likely our real inspiration. It is only after the desire arises there that it is passed to our orbitofrontal cortex, which creates our conscious experience of it.[26]

The wanting system is more fundamental than the system for liking. It is found in all species of animals, even the simplest and most primitive.[27] It evolved before the liking system, and in fact in the most ancient animals there is no liking system; wants are driven strictly by survival needs such as food and water. That's possible because creatures can survive if programmed to want whatever it is that they need, without ever having the experience of liking it.

If the reverse were true—if an organism were programmed to like what it needed without wanting it—it would not be motivated to fulfill its needs, and it would die. But the liking system present in the higher forms of animal life does serve a very useful purpose. It frees us from having our wants and desires *directly* trigger action. Instead, wanting is stimulated by liking, but not automatically. Before activating our wanting circuits, our brain takes liking into account, as well as other factors. For example, food is a basic need, and we are programmed to like it. Yet when we see a piece of alluring food, rather than mindlessly devouring it, we may pause while our brains balance the pleasure of eating with various nutritional and aesthetic considerations. It was the evolution of the liking system that gave animals the possibility of this more nuanced behavior, in which we may pass up something we are drawn to. It is interesting to note that because such "self-control" decisions are governed by our conscious mind, we can generally enhance that ability through practice and determination.

Recently, Berridge filled in another gap in the motivation picture. Reward system research had traditionally focused on the motivation to acquire things but not to avoid them, which seems equally important. Then, a few years ago, Berridge discovered that the accumbens governs not just wanting but also its opposite—the motivation to avoid or flee.[28] While one end of that structure creates desire, the other, apparently, originates dread. In between, there is a gradient. Berridge compares it to a musical keyboard that can play at both extremes but also hit many intermediate notes.

What is most interesting about this discovery is that the accumbens keyboard can be retuned by context and psychological factors. A stressful and overstimulating sensory environment such as excessively bright lights or loud music expands the border of the dread-generating zone while shrinking the border of the desire-generating zone. On the other hand, a quiet and comfy ambience alters the keyboard in the opposite manner, expanding desire and shrinking dread.

These are phenomena worth taking note of, for they can

operate on an unconscious level and exert their effects without you being aware of their root. I had a friend who worked in a noisy office and noticed that since taking the job, she always seemed to have an undercurrent of anxiety, though she could not pinpoint the problem at work. She finally suspected the noise and started wearing headphones, and the anxiety dissipated. Some people are more affected by such environmental factors than others, but in general Berridge's work helps explain why we might react differently to the same situation in different environmental contexts.

Through years of careful research, Berridge created a revolutionary new theory of the reward system. He had to struggle to do it. Roy Wise, his early mentor, did not accept his conclusions. Neither did anyone else. And so, for the first fifteen years Berridge had to work on his theory without any funding, fitting it in around other projects. Once he finally procured money in 2000, he was able to pick up the pace. But it still took another decade and a half for his ideas to catch on. Only recently have doubters drifted off; since 2014, his papers have drawn a steady four thousand citations each year. "Kent is one of the great pioneers," said his current Oxford collaborator Morten Kringelbach. "And [he] got there by ignoring what everybody told him."

OBESITY AND PROCESSED FOODS

In the latter stages of World War II, my father was a prisoner in the concentration camp Buchenwald, named for its location in the beech forests of Weimar, Germany. Though thousands of Buchenwald inmates died as a result of human experimentation, hanging, or shootings executed at the whim of the SS guards, the theory on which the camp was founded was called *Vernichtung durch Arbeit*, or "Extermination through labor." The plan was to work the prisoners to death.

My father was sent to Buchenwald in late 1943. His body weight became a clock ticking down toward his demise. One hundred and sixty-five pounds in his prime, by the spring of 1945 he weighed half that. Then, on April 4 of that year, the U.S. Eighty-Ninth Infantry Division overran Ohrdruf, an outlying subcamp of Buchenwald. In the days that followed, with the U.S. Army closing in, the main camp was evacuated by the Germans. Thousands of the prisoners were forced to join the evacuation "death marches." Others, however, were able to take advantage of the chaos. My father was among those. He and a friend, Moshe, headed deep into a cellar, where they hid behind a pile of boxes. There they huddled in the cold for several days, with only each other for warmth and no food or water, afraid to emerge.

On April 11, at 3:15 p.m., a detachment of troops of the U.S. Ninth Armored Infantry Battalion arrived at the Buchenwald entrance gate and liberated the camp. It was not a silent entrance, and my father and Moshe heard the commotion. Eventually, they emerged from their hiding place. Back in the light, they now encountered American soldiers, many in their teens or barely out of them, horrified at the sight of the emaciated prisoners and the corpses that were still piled everywhere.

The Americans were generous. They offered my father and Moshe whatever they had. Chocolate, salami, cigarettes, canteens of fresh water. Starved from both years of deprivation and days of total abstinence, my father later told me, they would have found even a rodent or puddle of water appealing. But on that day my father and his friend were offered quite a feast. My father restrained himself, but Moshe kept eating and eating. He consumed an entire salami. Within a few hours, Moshe had intense intestinal distress. He died the next day.

As in all aspects of our constitution, there are individual differences, and my father's makeup led him to restrain himself where poor Moshe didn't. In general, the mammalian motivational system is meant to operate within a range of ordinary circumstances, but not at the extremes. In extremis, we are all woefully challenged creatures. Rats, for example, when put on a restricted feeding schedule during which they receive a reduction in the amount of food that free-eating rats consume, will, when afterward allowed free access to food, gorge themselves just as Moshe did.[29] When the environment goes awry, our normally adequate neural systems can lead to our demise, as happened to my father's friend. That's a problem whenever society is disrupted, and it's an everyday problem for the victims of an out-of-balance or misguided reward system.

And then there are those whose job is to make sure that our reward system *is* misguided, because they profit from it. Consider the processed food industry. Around the turn of the millennium, a purveyor of packaged frozen cheesecake temporarily revived its longtime slogan, "Nobody doesn't like Sara Lee."[30] Ten years later, the neuroscientists Paul Johnson and Paul Kenny noted just how right they were, pointing out that the set of those who appreciate Sara Lee includes the mice and rats that scientists study. It is doubtful that the Sara Lee Corporation would ever use "Even rodents love Sara Lee" as their slogan, but there is good reason for their products' universal allure: it is a symphony of sugar, fat, salt, and chemicals orchestrated to please but never satiate.[31] The mix is so addictive and unhealthy that when Johnson and Kenny offered the rats cheesecake alongside their usual chow, they ballooned in weight from 325 to 500 grams in just forty days and showed pathological changes in parts of their brain. That's pretty impressive, even for the complex thirty-ingredient chemistry-lab-in-a-box that is Sara Lee.[32]

To be fair, the rats in the study loved not just Sara Lee but many other highly processed foods as well. Given access to a twenty-four-hour "cafeteria" that also included frosting, candy, and pound cake, the rats were subjects in a laboratory experiment on diet and the reward system. The researchers' aim was to study the "addiction-like reward dysfunction" that leads to compulsive eating. Inducing compulsive eating through junk food proved alarmingly easy because inducing it is precisely the goal of most processed food and fast-food purveyors. As the former Coca-Cola executive Todd Putnam said, his marketing division's efforts boiled down to: "How can we drive more ounces into more bodies more often?"[33]

It might seem odd to refer to people's overconsumption of

processed foods as an addiction, but the term today is not as narrowly defined as it has been in the past—as a chemical addiction, such as is associated with drugs and/or alcohol. Instead, based on new neuroscientific research, addiction has come to be understood in a far broader sense. Today gambling, internet use, gaming, sexual acting out, and food are all considered possible subjects of addiction, with a common root. To reflect that, the American Society of Addiction Medicine redefined addiction in 2011 as "a primary, chronic disease of brain reward."[34]

When our reward system is operating as evolution meant it to, liking and wanting act in tandem, albeit in a nuanced, complex manner that allows us to distinguish between them. If we like sex, or ice cream, we may be motivated to pursue it—or as Berridge showed, we may not. But addictive substances and activities cause physical alterations in the accumbens, dramatically increasing the amount of dopamine released and overstimulating the organism's wanting circuits.[35] Each episode amplifies that effect, producing ever-stronger urges to repeat the addictive behavior. Scientists call that "sensitization." The physical changes are long lasting and may even become permanent. Sadly, addictive drugs often have the opposite effect on the liking system. The subjective pleasurable effects of the drug are decreased due to the development of tolerance. As a result, the longer the addiction, the more the drug is wanted and the less it is liked.

Some people are especially vulnerable to this dynamic. Geneticists employing new technologies are uncovering a genetic connection; an individual's susceptibility to addiction seems to depend upon genes related to the dopamine receptors in the wanting system.[36] Because addiction of one sort or another is fairly common, you might think it indicates a major flaw in our genetic design. But it doesn't really; addiction is rarely found in natural settings. It was not an issue in nomadic societies of hunters and gatherers, and rats and mice suffer from it only when exposed to human creations in a laboratory setting. Humans today suffer from addiction only as a by-product of "civilized" human society, in which we create thirty-ingredient cheesecakes, dangerous drugs, and other products that the Nobel Prize–winning scientist Nikolaas Tinbergen called "supernormal stimuli."[37]

ADDICTION AND SUPERNORMAL STIMULI

Tinbergen stumbled upon the concept of supernormal stimuli in an unlikely setting—while studying stickleback fish kept in tanks in his lab in Holland. Male sticklebacks have a bright red underbelly. Even when kept in an aquarium, they stake out a territory and attack other males that enter it. To study that behavior, Tinbergen and his students guided dead fish on wires toward the male defenders. For convenience, they later replaced these with wooden facsimiles. They soon realized that it was the redness of the underbelly that provoked the attacks. Sticklebacks would not bother an accurately shaped wooden dummy if its belly wasn't red, but they would assault very unfishlike objects that did have red undersides. Males kept in tanks by the window would even enter attack mode when a red van drove past. Most important, Tinbergen noted that the fish would ignore a real stickleback in order to fight a dummy, if the dummy was painted a brighter red than the real fish.

That bright red dummy fish was a supernormal stimulus: an artificial construct that stimulates an animal more powerfully than any natural stimulus. Tinbergen discovered that it wasn't hard to create such stimuli. A goose accustomed to rolling a stray egg back into the nest would ignore her egg and instead attempt to retrieve a much larger volleyball. Hatchlings would ignore their parents and seek food from a fake beak mounted on a stick if the beak had more dramatic markings than their parents'. Everywhere Tinbergen looked in the animal kingdom, it seemed possible to hijack an animal's natural behavior if one employed an artificial stimulus that was strategically constructed for exaggerated appeal. And that's just what the processed food purveyors, the tobacco industry, the illicit drug cartels, and, in the case of opioids, Big Pharma are doing to their human "customers."

Most addictive substances and activities are supernormal stimuli, and they disrupt the natural balance of our personal world just as they do that of the stickleback-fish world. For example, most addictive drugs are plant materials that have been refined into highly concentrated substances, then processed to make them more potent and allow the active ingredients to be absorbed into the bloodstream faster.[38]

Consider, for example, the coca leaf: When it is chewed or stewed as tea, it produces only mild stimulation and has little addictive potential. But when refined into cocaine or crack, the drug is quickly absorbed and highly addictive. Similarly, we wouldn't have an opioid epidemic if the only way to consume opioids was to chew on a poppy plant. The same goes for cigarettes: as the tobacco plant is harvested and processed into a form that can be inhaled as smoke, hundreds of additional ingredients are added to enhance its flavor and aroma and to speed absorption into the lungs, making the resulting tobacco product significantly more addictive than unprocessed tobacco. Alcohol, too, is the product of processing. If, rather than buying vodka at the store, a person had to rely on eating naturally occurring rotten, fermented potatoes, there would be very few alcoholics.

The obesity epidemic also has its roots in supernormal stimuli, or, as scientists in the food realm call them, hyperpalatable foods. To avoid malnutrition, our brains evolved to like calorically dense foods, foods such as berries and animal flesh that are high in sugar and/or fat, but these were relatively scarce, and obesity was rare. In the preindustrial era humans survived on an unprocessed diet high in protein, grains, and produce and relatively low in salt, and obesity remained rare. In the last several decades, however, commercial food processors have learned to alter food in a manner analogous to the processing drug dealers use to create addictive drugs. Once they discover what our reward system responds to, they deliver it in an unnaturally concentrated form, and one that is rapidly absorbed into the bloodstream; just as with illicit drugs, both the concentrated nature of the food substance and its rapid absorption into the bloodstream increase the reward system effect.

Today food companies spend millions of dollars researching how to create these hyperpalatable foods. They call it "food optimization." Said one Harvard-trained experimental psychologist who works in the field, "I've optimized pizzas. I've optimized salad dressings and pickles. In this field, I'm a game changer."[39]

Food optimizers are game changers because hyperpalatable food can interfere with a person's natural tendencies, just as the volleyball interfered with the goose's maternal instinct, or the fake beak did with the hatchlings' feeding. The result is that people crave the optimized food far more than is warranted by the pleasure it delivers.

Obesity has been estimated to cause 300,000 deaths per year in the United States alone.[40] It is a situation that developed gradually, so that, like the proverbial frog in a pot of water that is heated ever so slowly, we failed to notice what was happening until it was too late. The availability of drugs of abuse and the advance of commercial food science have both contributed to fooling the human emotional reward system. Science can elucidate the mechanisms through which foods addict us, but it's up to consumers to heed the warning and avoid being manipulated into obesity.

The design and mechanics of our wanting and liking system are fascinating, as is the story of how we discovered its mechanisms. Now that we know on a molecular level how our reward system works, some learn to manipulate it for profit through our behavior and biochemistry—just as the tobacco, food, and drug manufacturers (both the cartels and in some cases Big Pharma) have done. But as educated consumers, we can use our knowledge of what they are doing to defeat their ends by making better, healthier choices.

7

Determination

Mike Tyson stood close to his opponent. It was the world heavy-weight championship in Tokyo, Japan, in February 1990. There were five seconds left in the eighth round.[1] Tyson's opponent, James "Buster" Douglas, was never supposed to have made it this far. Now, with his elbows out and his gloves meeting in front of his chin, Douglas's arms formed a tight ring. Douglas stared down at Tyson, who, with his knees bent, seemed a head shorter. Douglas appeared to be baiting him.

In a flash Tyson straightened. His right glove shot up inside the ring of Douglas's arms, a fierce uppercut that landed squarely under Douglas's chin. Douglas's head snapped to the right. His legs buckled. He staggered backward, fell hard onto his backside, and slid two feet on the canvas.

Douglas was dazed as the referee began his count. The referee got to seven before Douglas finally leaned on his elbow and started to push himself to his feet. By the count of nine he was up but wobbling. "[Knocked] out on his feet," as the HBO TV analyst Larry Merchant described it. Had this happened ten seconds earlier, Tyson would have moved in and finished Douglas off, but the round ended, and Douglas was saved by the bell. He made it to his corner, where he had sixty seconds to shake off the cobwebs before the next round.

Just before the opening bell, Merchant had said a Douglas win would "make the shocks" in Eastern Europe—which was going through the chaos that led to the fall of the Soviet empire—"seem like local ward politics." His fellow analyst, Sugar Ray Leonard, said it would "shock" the world if Douglas even made it past the first few rounds. In Las Vegas, the Mirage bookmaker Jimmy Vaccaro had opened the betting with odds favoring Tyson twenty-seven to one. Still, "people jumped on Tyson," he said. Attempting to balance the betting, he raised the odds to thirty-two to one, and eventually to forty-two to one. No other casino even offered action on the match; they couldn't find anyone willing to bet on Douglas. Instead, they offered bets on how long the fight would go—that is, on how long Douglas would last before Tyson felled him. Tyson had knocked out his opponents in his last five title fights. His prior opponent had lasted just ninety-three seconds.

Douglas wasn't originally scheduled to fight Tyson. The "real" fight everyone was waiting for was a battle in Atlantic City the following June, between Tyson and a more accomplished boxer, Evander Holyfield. In fact, at a dinner the night before the Douglas match, the boxing promoter Don King, the then casino mogul Donald Trump, and Shelly Finkel—Holyfield's manager—had met to discuss plans for the Atlantic City fight, for which Tyson was guaranteed $22 million, and Holyfield, $11 million. No one cared about the Douglas fight. The Tokyo match was just a warm-up, added late, a chance for the champion to scoop up some extra cash while awaiting the big event. For Tokyo, Tyson was paid $6 million. To take his beating, Douglas, the nobody, was offered $1.3 million. It was far more than he'd ever previously been paid.

If no one else was impressed with Buster Douglas, his mother was. As Douglas was training for the fight, Lula Pearl Douglas started bragging around town about her son. He asked her to rein it in, but she didn't. He was going to fight the champ, she'd say, and he was going to "kick his ass." Douglas, too, fantasized about upsetting everyone's expectations and about all the things he could buy for his mother with his winnings.

Three weeks before the fight, Douglas was awakened by a phone call at 4:00 a.m. His mother had suffered a massive stroke.

She'd died almost immediately. She was forty-seven years old. Douglas was devastated. "I was just in a shell," he said. "No one could understand my plight. I lost my best friend, my mother. I had really no one to turn to." His handlers offered him a chance to back out. He didn't. "She would have wanted me to stay strong," he said.

Of Douglas's eighth-round knockdown, *The New York Times*'s James Sterngold said, "My immediate reaction was, This is over." That was the reaction of most people: when Mike Tyson knocks you down, you don't get up. Back in his corner, Douglas knew that if he did climb back into the ring, Tyson would come after him savagely, looking for KO number thirty-four. But Douglas didn't have to continue the fight. No one had expected him to get as far as he had, or to get up from the punch he'd taken, and no one would have blamed him if he'd decided to take his $1.3 million and end it there. Yet he chose not to do that. He stood up and went out to face Tyson again. Two rounds later, with a minute and fifty-two seconds remaining in the tenth, Douglas landed a barrage of punches that knocked Mike Tyson out. Today, decades later, it still stands as the greatest upset in the history of boxing.

After Douglas beat Mike Tyson, others started to beat him. Tyson was a finesse fighter, but enormously aggressive and powerful. Boxers had been afraid of him. But Douglas showed that if you had the grit to withstand the first few rounds, he'd start to tire, and then the game would be different. His aura stripped away, Tyson was never the same. Douglas, too, soon faded. He ended up fighting Holyfield in Tyson's place, and got the $20 million payday, but his heart was no longer in it. He was knocked out in round three and retired soon after.

When, after the Tyson fight, an interviewer asked Douglas how he did it, how he could have come back from the round-eight knockdown and keep attacking, how he, the no-name, could have done what no one had ever done, and knocked out Mike Tyson, Douglas teared up. "My mother," he said. "My mother . . . God bless her heart." She had believed in him, and he was driven to live up to her dream. It was a touching, if clichéd, moment, but it sheds light on one of the most important factors in human experience—determination. That night in Tokyo, Douglas had far more of it than Tyson, far more, too, than Douglas would have in his later fight with Holyfield.

If you had asked Muhammad Ali how many push-ups he could do, he'd say, "Nine or ten." He could obviously do many more than that, but, as he wrote in his autobiography, he didn't start counting until he'd done so many that it hurt like hell to keep going.[2] Buster Douglas didn't have Ali's grit, but for one match the death of his mother had activated an iron determination to win.

We run into many barriers on our way to achieving goals. Limited talent, financial troubles, situational and physical issues, may get in our way. But determination is a tool that can shatter those barriers. That is true in all life's contexts, but it is especially apparent in sports, with their fixed rules, clear winners and losers, and definitive statistics. In fact, the Douglas victory was far from unique: over and over again, throughout sports history, we've seen extraordinarily determined human spirits achieve what others thought impossible. To run a 4-minute mile, for example, was a feat athletes had pursued for decades, without success. The human body, experts said, was incapable of accomplishing that, and athletes were warned that it could be dangerous to try. Then, on May 6, 1954, the medical student Roger Bannister ran the mile in 3:59.4. A month later the Australian John Landy clocked in at 3:58. Soon, for a top runner to break the 4-minute mark became routine. According to *Track & Field News*, about five hundred Americans have now broken that barrier, and a couple dozen are added to the list each year.[3] It is as if a switch were thrown. Not a physical switch, a mental switch—the realization that the task could be done, which led to the determination to keep pushing until one accomplished it.

Shakespeare asked, "Whether 'tis nobler in the mind to suffer the slings and arrows of outrageous fortune or to take arms against a sea of troubles, and by opposing end them?"[4] The answer nature provides organisms is clear: take arms against your troubles.

In the preceding chapter, we looked at motivation—the reasons (wanting and/or liking) that we have for acting in a particular way. In this chapter, we'll examine the related issue of determination— our firmness of purpose in pursuing the goals we are motivated to achieve, in spite of obstacles and challenges. One can debate the evolutionary origins of our feelings, and the nuances and purpose of all the emotions a human experiences, but a powerful lesson of the new science of emotion is that on the most primal level one purpose of emotion—not just in humans, but in other animals, too, including the lowliest—is to provide the psychological resource to embrace opportunity and to face, endure, and overcome challenges. Amazingly, scientists now understand the origin of determination. They can pinpoint precise circuits in your brain that, when damaged through disease or injury, will leave you listless, but, when fired up, will put you in the same mode as Buster Douglas the night he beat Tyson.

WHERE DETERMINATION COMES FROM

In June 1957, Armando, a fourteen-year-old boy from Chile, was awakened by a severe headache that persisted for about fifteen minutes.[5] The episode left no lingering trace. But a few weeks later he experienced another such episode, this time during his waking hours. After a third incident Armando's doctor advised that his parents take him to the Mayo Clinic. There, tests revealed a small tumor in one of the fluid-filled cavities, or ventricles, near the midline in his brain. The tumor was surgically removed in early August.

Before the surgery, Armando was a pleasant youth of normal demeanor and average intelligence. After the surgery, he became completely indifferent to his environment. He did not move his eyes to gaze about the room or engage in any voluntary motion. If put in a clearly awkward position, he'd make no effort to move into a more comfortable one. On command he would grasp an object firmly, but without a word or other reaction. He would not speak unless spoken to, and even then he would answer in terse replies. He made no effort to eat, and if food was placed in his mouth, he would swallow it whole without chewing or reacting in any way to its taste. He could identify his parents but showed no emotional reaction to them, or to anything else. If there ever was a diametrical opposite to the determination of Buster Douglas, it was the profound apathy embodied in this boy.

About a month after the operation, as the swelling of Armando's brain began to subside, his apathy faded. He began to react to his environment, to pursue goals, to interact with those around him. His former nature, rather suddenly, began to return. He called out his parents' names and once more spoke spontaneously. He started to greet the doctors in a friendly fashion, to laugh at jokes, to show interest in his surroundings. He strove so diligently to learn English that it soon became possible for the non-Spanish-speaking staff to converse with him in simple sentences. At the time, no one understood why this happened. What brain structures could the swelling have interfered with? Research done fifty years later would reveal a likely explanation.

As a species, we have a prime directive to survive and reproduce. But we also have secondary programming granting us the determination to seek reward and avoid punishment. Determination is a trait provided to us by evolution because it supports our prime directive, and like all mental phenomena it has both a psychological and a physical component; the Buster Douglas story illustrates the former, Armando's the latter. The two are deeply intertwined, so though determination arises from a process in your physical brain, it can be accessed through psychological events. Losing a loved one changes your brain. So does a pep talk. So does brain surgery. And as we'll see, in the longer term so do exercise and meditation.

Most emotional processes are distributed over many brain areas in a complicated manner. We've seen that the source of wanting and liking is our reward system. Determination, too, is a complex and multifaceted mental phenomenon, and until recently neuroscientists did not expect to be able to pinpoint a well-defined network or pathway directly involved in creating it. It therefore came as quite a shock when a cabal of neural circuits that govern the physical side of determination was discovered in 2007.[6] It comprises two networks that are distinct but work together, the "emotional salience network" and the "executive control network."

The emotional salience network consists of tiny nodes that are anchored in a set of structures that were previously associated with a variety of roles in our emotional lives. These are the so-called limbic structures, such as the insula, the anterior cingulate cortex, and the amygdala, which I mentioned in the introduction. The executive control network, in contrast, includes sites in the executive prefrontal cortex, a region known to have a role in sustained attention and working memory.

When the steady stream of new high-tech methods for studying the brain began flowing in the 1990s, it appeared that we'd soon be able to clarify the function of all the gross brain organs the anatomists had long ago identified. But that didn't happen. Instead of providing clarification, the new technologies revealed such a dazzling degree of complexity that it took a long time for scientists to begin to comprehend what they were looking at. One of the surprises was the number of finer substructures and distinct regions within the gross structures. In addition, the neural wiring connecting all these structures proved to be enormously complex. Those who endeavored to trace it produced maps that looked more like a pot of spaghetti than the simple circuit diagrams they had hoped for.

These recent discoveries support the view that few (if any) brain functions are localized. One might create a specific effect by stimulating or destroying a localized bit of brain tissue, but chances are, the tissue in question is but a cog in a much larger apparatus: general brain function usually occurs through the interactions of multiple networks of nodes—some large, some as tiny as a few millimeters in diameter—that are scattered throughout multiple brain structures. The emotional salience network and the executive control network are two such sets of anatomical structures.

The word "salient" means "most noticeable or important,"

and it describes what the salience network does, which is to monitor our internal emotions and our external environment and take note of what is important. "Our brain is constantly bombarded by sensory information, and we have to score all that information in terms of how personally relevant it is for guiding our behavior," said William Seeley, a neurologist at the University of California, San Francisco and one of the scientists who discovered the network.[7] The emotional salience network identifies the most relevant among those inputs and spurs you to act (or not act) on that basis.

The executive control network has the job of keeping you focused on what is relevant to your goals while ignoring distractions. It jumps into action once the salience network has been activated. It then marshals the brain's resources to enable you to act if necessary.

Just how it feels when a node in your salience network is stimulated was vividly illustrated—by accident—in 2013, by a team of neurologists at Stanford Medical School who were trying to pinpoint the source of seizures in a patient with severe epilepsy.[8] They hoped to remove the offending tissue, but only if the excision would not hinder the patient's well-being. To identify the problem area, they implanted electrodes in various areas of the man's brain, applied a few milliamps of current, and observed the physical response. They also asked him about his sensations, thoughts, and feelings.

On one of the implantation cycles, the man's answer shocked them. He reported feeling "determination." The feeling wasn't associated with any particular goal; it was just an abstract feeling. He likened it to the emotion you'd have if you had to drive up a hill into an ominous storm. Not dread, but a positive feeling that says, "Push harder, push harder, push harder to try to get through this." Just as Buster Douglas experienced, except, the patient emphasized, there was no vision of any particular challenge to conquer, just a feeling of determination, devoid of context.

The doctors had gotten lucky. They had inadvertently implanted the electrode in a single tiny node of the complex

salience network. When they moved the electrode a few millimeters one way or the other, the effect did not occur, but when it
was planted just right, the patient reported an urgent sense of the
need to act, or to persevere. The doctors then looked for the node
in a second patient and found it in the same anatomical position.

"Our study pinpoints the precise anatomical coordinates . . .
that support complex psychological and behavioral states associated with perseverance," said the neurologist Josef Parvizi, lead
author of the study.[9] Though they had stimulated just one node
in a large network, and therefore produced a feeling unattached
to any specific goal or context, he marveled that "electrical pulses
delivered to a population of brain cells in conscious human individuals give rise to such a high level set of emotions and thoughts
we associate with a human virtue such as perseverance."

The important role played by the salience network is reflected
by its multitude of nodes and their wealth of connections to other
parts of the brain. Sitting along the midline of that organ, the
salience network engages in dialogue with both the executive control network and other "executive" areas of the frontal lobe, as
well as subcortical parts of the brain that are involved in complex
emotions and in producing physiological reactions. As a result it is
informed by both what we think and what we feel.

When the effect of salient stimuli is quieted by, say, the taking of a beta-blocker, a patient's verve is diminished, leading to
unusually sluggish reactions.[10] When elements of the network are
badly disrupted, as in the case of Armando, the result is profound
apathy. But when they are supercharged, the result is profound
determination. You desperately feel "the need to act, the need to
endure," said Seeley. That seems to be the condition induced in
Buster Douglas by the death of his mother; it was as if, within his
brain, a "determination" switch were turned on.

It probably seems simplistic to imagine that an episode of iron
will could be traced to a specific and discrete process within the
brain and that modern technology is powerful enough to identify that process. But in an astonishing 2017 experiment, scientists
demonstrated the power of that discovery by creating their own

Buster Douglas miracle in a laboratory setting: they used brain stimulation to activate a kind of "grit switch" in the emotional salience/executive control complex.[11]

The "champion" and the "challenger" in the 2017 experiment were not boxers but rodents engaged in what one might think of as a mouse analogue of the Douglas/Tyson fight. You can't force mice to box each other, but you can force them to engage in physical competition. To accomplish that, the scientists released two of the creatures into a narrow tube, one at each end. The natural instinct of both mice was to get out by pushing forward. But because the tube was so narrow, only one of the two mice could do that. The other would have to give up on that strategy and instead back out, so this was like a reverse tug-of-war. Each mouse initially tried to move forward, but one would eventually retreat. They were of comparable stature, so these were tests of determination, not physical strength.

The experimenters ran a series of such contests, determining the winners and losers. They then gathered the losers and, employing a cutting-edge technique called optogenetics, arranged so that they could stimulate each mouse's determination switch at will. Could they use that switch to turn the vanquished into victors?

The method involved sending a pulse of laser light through a cable they had implanted at a precise point in each mouse's brain. In that way, they were able to turn nearby neurons either "on" or "off." Having prepared the losing mice in that manner, the researchers arranged a series of rematches against the former winners. This time, with their determination switches on, 80 to 90 percent of the prior losers won their rematch.

Personally, I find the mouse version of this saga as inspiring as the Buster Douglas version. The Douglas tale suggests that with the proper emotional determination, we can boost ourselves to become superhuman (where "human" means our old selves). But the mouse story assures me that that belief isn't just wishful thinking: by stimulating the right handful of neurons, we really can increase resilience and determination.

The Stanford group's findings suggest that inborn differences in structure and function of the executive control network are linked to a person's ability to cope with difficult situations. "These innate differences might potentially be identified in childhood," said Parvizi, "and be modified by behavioral therapy, medication, or, as suggested here, electrical stimulation." In the years since Parvizi's work, there has been much research on those and other ways to enhance the network, and fortunately we don't have to have a relative die or shine a pulse of laser light into our brain to do it.

Two techniques stand out. One method, if you're sedentary, is through aerobic exercise. Recent research concludes that exercising as little as fifteen minutes per day to improve your cardiac fitness leads to better executive control function.[12] That may seem like an improbable link, but exercise has been shown to increase a "growth factor" called BDNF. Growth factors are like a kind of fertilizer for your brain. They help it create new connections, which, in general, increases its computational power and is what the brain needs to learn and adapt. In animal studies, increased levels of BDNF have been found to reduce depression and enhance resilience. Of course, to add exercise to one's life could be an uphill battle, for if you have poor executive control, you may not have the determination to do the exercise. But if you can make yourself start, then the ever-increasing executive function that results from your exercise will make the decision to engage in exercise ever easier, leading to a positive feedback cycle in your favor.

Another way to increase determination is through mindfulness meditation, which teaches attention control, emotion regulation, and increased self-awareness. In one study, two weeks of mindfulness training in a group of smokers led to a 60 percent reduction in smoking, a very difficult achievement.[13] Brain imaging after the meditation program confirmed a significant increase in activity in the executive control network.

Some people have naturally high executive control and are born "doers." They let nothing get in the way of their goals. For them, determination is a way of life. For the rest of us, it's nice to know there are things we can do to enhance it.

THE APATHY OF COMPUTERS

That our emotion system has the power to determine when we must act is one of the features that distinguishes humans—and other animals—from computers. Consider, for example, the robot Sophia, developed in 2015 by Hanson Robotics. With her humanoid face modeled after the actress Audrey Hepburn, Sophia looks human, sounds human, and is capable of impressive facial expressions. But she isn't really humanlike.

Sophia was programmed to react in a particular way to each of a large number of stimuli. Her conversation ability, for example, comes from sets of canned responses that are included in her programming. Despite her impressive appearance, like other computers, she is incapable of independent thought and action. If you took her to a room, or a garden, or the middle of a busy street and turned her on, what would she do? Would she explore the room? No, that would take curiosity. Would she gaze at the beauty of the garden? No, she has no sense of enjoyment. Would she carefully walk off the street and onto the safety of the sidewalk? No, she has no drive to avoid harm.

A robot like Sophia can be charming. She can produce humorous lines and disarm you with her easy banter. But she does not "decide" to act in the sense that a human does. She merely executes a fixed set of instructions, beginning when she is activated and ending when her program reaches the instruction to stop. And so, if, during one of her public appearances, something occurs that her programmer hadn't planned on, she would not react. If a fire alarm went off, she would not flee. If offered a piece of chocolate, she would not be tempted.

To execute a set of pre-scripted behaviors in reaction to a predefined trigger is a skill that developed early in our evolutionary history and remains in the playbook of all animals, including us humans. But higher animals, unlike more primitive life-forms—and Sophia the Robot—also possess the ability to *decide* whether to act based on their assessment of a *novel* situation, one that presents no preordained trigger. This ability is actuated through processes of ever-increasing levels of complexity. As we saw in chapter 3, at the most primal echelon, there is a capacity for feeling that is binary—dividing all experience into either good or bad—which psychologists call core affect. Then there are the basic feelings of fear, anxiety, sadness, hunger, pain, and so on. And in humans the brain circuits also produce nuanced and sophisticated social emotions such as pride, embarrassment, guilt, and jealousy. Ultimately, the interaction of all these levels of emotion produces an urge to act, or to refrain from acting: to decide whether to take action, and to motivate us to keep trying even if the quest is difficult or unpleasant, is one of the great gifts of our emotions.

THE DETERMINATION TEST

The best way to understand the role of determination in the animal brain is to recognize that navigating the environment requires a constant assessment of the costs and benefits of potential actions. Our determination circuits help decide how important our goals are, which possible actions deserve further attention or action, and which should be ignored. Those circuits feed into our thought, sensory, and motor circuitry, altering those neural processes to make them more efficient. Whatever task we are confronting, whatever problem we are trying to solve, our inherent mental and physical ability will be greater if we are motivated to succeed.

A person's degree of determination at any given moment will, of course, depend on the situation. But each of us has a baseline level of determination or drive, and psychologists have developed a questionnaire to assess it.[14] It can be filled out by a therapist, or simply someone who knows the subject well. It can also be self-administered—that's the version I give below—though in patients with significant mental impairment, a more accurate assessment is obtained if the questions are answered by a friend or family member.

A scientist who studies motivation wrote, "Apathetic . . . non-optimal human functioning can be observed not only in our psychological clinics but also among the millions who, for hours a day, sit passively before their televisions, stare blankly from the back of their classrooms, or wait listlessly for the weekend as they go about their jobs."[15] How determined are you? To take the test, answer each question as either "very true," "somewhat true," "slightly true," or "not at all true."

1 = very true 2 = somewhat true 3 = slightly true
4 = not at all true

___ 1. I am interested in things.
___ 2. I get things done during the day.
___ 3. Getting things started on my own is important to me.
___ 4. I am interested in having new experiences.
___ 5. I am interested in learning new things.
___ 6. I put a lot of effort into things.
___ 7. I approach life with intensity.
___ 8. Seeing a job through to the end is important to me.
___ 9. I spend time doing things that interest me.
___ 10. No one has to tell me what to do each day.
___ 11. I am as concerned about my problems as I should be.
___ 12. I have friends.
___ 13. Getting together with friends is important to me.
___ 14. When something good happens, I get excited.
___ 15. I have an accurate understanding of my problems.
___ 16. Getting things done during the day is important to me.
___ 17. I have initiative.
___ 18. I have motivation.
____ TOTAL

In this inventory, a high score indicates general apathy, a low score determination. The lowest possible score is 18. The average score of healthy young adults is 24. For those in their sixties, the average rises to 28. About half of all subjects score within four points of the average for their age-groups, and two-thirds score within six points.

The inventory was designed to measure your baseline level of determination, which is usually stable over time as opposed to the ephemeral determination you may feel having to do with immediate circumstances. On the other hand, as I mentioned, it is possible to increase your baseline determination level employing long-term techniques such as exercise and meditation. Also, certain illnesses can erode your baseline level. In fact, the scale was developed to assess not healthy individuals but those suffering from issues that diminish determination such as traumatic brain injury, depression, and Alzheimer's disease. The average is 37 for those aged thirty to fifty who suffer from traumatic brain injury, 42 for people suffering from depression, and 49 for those with moderate Alzheimer's.

DECLINING WILL

In extreme cases, such as in patients suffering from a condition called fronto-temporal dementia, the degradation of the emotional salience network can lead to severe apathy, like that which affected Armando. Armando's apathy came on suddenly as the result of the post-surgery swelling of his brain, which did temporary damage. By contrast, because dementia occurs gradually, it may offer a unique opportunity to observe, step-by-step, over a long period of time, how behavior changes as the emotional salience network is slowly muted.

That was what I saw happening with my mother. When I was a child, my parents, my two brothers, and I lived in a tiny walk-up apartment. The close quarters would have been easier to tolerate had a third of our living space not been off-limits. The do-not-enter zone was the living room/dining area complex that my parents furnished with items a couple of steps above the cut-rate furniture in the rest of the apartment. There, the floor was carpeted, the table protected by a felt pad, and the couch and easy chairs covered with plastic sheaths that my high school friends referred to as furniture condoms. The border was not blocked with anything physical, but the "keep out" order was as potent as it would have been had the area been marked off with police tape.

The off-limits area was used only on Passover and the days that we Jews call the High Holidays. When I was growing up, the Jewish holidays meant you got to miss school, had to pray at the temple, and, at dinner, would be having the standard kosher brisket and potatoes. To enter the forbidden room at any time other than the eighteen days encompassed by Passover and the period between Rosh Hashanah and Yom Kippur would bring a fierce yelp from my watchdog mother. I've shared her attitude about that living space not because it was unusual but because that zeal was typical of her. If my mother deemed the floors dirty, she wouldn't reach for a broom or mop; she'd get on her knees and scrub it as if it were an operating room. When she was older, if, despite her sore and swollen knees, she took a walk, it wasn't around the block; it was for a couple miles. If she saw a politician she didn't like on the television, she wouldn't shake her head in disapproval; she'd mutter in Yiddish that cholera should befall him. And when I was a child and she told me she loved me, she didn't give me a peck; she blanketed my forehead with kisses. My mother was not apathetic about anything. I supposed her zeal a quality she'd needed to survive the camp the Nazis sent her to, where nine out of ten inmates perished.

After my father died, my mother moved to California to live in the guesthouse adjacent to my home. She was eighty and brought her then-fifty-year-old living room/dining room furniture with her. After a few years, I began to notice changes in her that I at first attributed to the mellowing of age. She became less concerned about our sitting on the forbidden couch on non-holidays. She agreed when I finally suggested stripping the fifty-year-old plastic covers off the furniture. With time, I realized that the change in her was something different from the smoothing of sharp edges. This was not mellowing. Nor did it seem to be depression—it didn't come with any of the usual symptoms such as sadness, hopelessness, or emotional distress. But there was an onset of a lack of caring that over the years would grow like a blight on her personality, gradually subduing it. Then, one holiday morning, the mother who used to scold me for being tardy at religious services appeared at her door in pajamas, indifferent about attending the High Holiday services I was going to take her to. I knew there was a problem.

Within a few years my mother needed enough help with everyday life that I moved her to an assisted living center. One morning soon after the transition, I stopped by the center's dining room to seek her out and share a cup of coffee. I found her at a group table, eating bacon and eggs. For reasons of Jewish dietary law, my mother had never eaten pork in her life. She'd been so adamant about it that I would have been less shocked if I had found her sitting at the table in her underwear. And so I stared. "What?" she said. At a loss for words, I stated the obvious. "Mom, you're eating bacon." She just shrugged and replied, "That's what they gave me. I like it." After a couple of years, she got to the point where, when left on her own, she would simply sit in a chair all day and stare at the television. It was time for her to move to a home with a higher level of care.

It was a long slow slide, but my opinionated, histrionic mother finally plateaued. As I write this, if you ask her what she wants to do, she will just smile. If you ask her what she wants to eat, she will just shrug. But if you place food in front of her, she will generally begin to eat, especially if you cut her a piece to get her started. Unlike Armando, she will place the food in her mouth, chew and enjoy it, and then continue to eat. And, happily, she can still carry on a simple conversation—if you initiate it. Her new "Don't worry! Be happy!" attitude is in some ways a refreshing improvement over her former "Do worry! Make yourself miserable!" but it is also sad for the internal decay that it signals.

The study of motivational changes in age-related cognitive decline is interesting to scientists because it helps connect brain structure and function. For the rest of us it is valuable because it encourages us to be aware of the alterations brought about by aging, and to do what we can to try to stave off mental decline by engaging in good health practices.

In addition to aging, there is one other life condition that, though not an injury or illness, has an important negative effect on determination: sleep deprivation.[16] Ever notice that when you lack sleep, things that seemed important before no longer seem to matter so much? Setting up and programming the coffeemaker so the coffee will be ready when I wake up the next day seems like a great idea at 9:00 p.m., but at 2:00 a.m. I decide it's not essential and I can prepare the coffee when I wake up. I notice the same phenomenon in my work. Normally, when I read through a chapter I've just completed, I find numerous rough spots I feel compelled to smooth out. But if I do that same exercise very late at night, the flaws don't seem important, and I fool myself into thinking that what I've written is great—at least until I look at it again after a good night's sleep. Because I realize that, I don't edit my writing when I'm in need of sleep.

Having proper sleep is crucial to maintaining motivation and, more generally, to our emotional health. For example, neuroimaging studies reveal significant activity during REM sleep in each of the structures that houses nodes of the emotional salience network. Experiments suggest that all that activity is related to an overnight resetting function within those key regions. One researcher asked twenty-nine healthy subjects to keep a detailed log of their activities and feelings over a two-week period.[17] He also asked them to keep a journal of their dreams. He found that between a third and a half of the emotional concerns that participants reported during the day resurfaced in their dreams that night—a large percentage considering that most dreams probably went unremembered. That's strong evidence that our sleep provides a nightly recalibration that restores the appropriate emotional salience responses that are necessary to guide appropriate decisions and actions.

So what happens if we don't get enough sleep? Plenty. For example, one study found that a single night of sleep deprivation triggers a 60 percent amplification in reactivity of the amygdala (assessed using fMRI) in response to emotionally negative pictures. A related study found that a night of sleep loss increases subjective reports of stress, anxiety, and anger in response to low-stress situations. Insufficient sleep has also been linked to aggression. And restricting sleep to five hours per night across an entire week was found to produce a progressive increase in emotional disturbance, such as exaggerated fear and anxiety (assessed on the basis of questionnaires and diary documentation).

What scientists have learned about determination and its opposite, apathy, gives us insight into a most basic function of our emotions. More fundamental in its effects than love or hate, happiness or sadness, or even fear and anxiety, determination is what gives us the urge to act—to reach out toward something or someone, to speak, or even just to move—and the energy to see our actions through until we achieve our goal.

As emotional beings, we have desires. We formulate the goals and subgoals that are dictated by them, from the sublime intent to write a novel to the insignificant objective of brushing our teeth. But before we can achieve any goal, whether great or small, we must be determined to act, and that is the role of our emotional salience network.

At our best, we human beings are energetic, spirited, and self-motivated. We expend effort, take action, exhibit commitment. That we have a self-starter capability and a good dose of perseverance is one of the signs that marks us as being alive. And it is present not just in us but in the most primitive of animals. For even a lowly fruit fly brain does not need to be told what to do. It knows how to make choices that enable it to avoid predators, pursue a mate, and comfort itself with alcohol when its overtures are rejected.

PART III

Emotional Tendencies and Control

8

Your Emotional Profile

"Every person is unique," Gregory Cohen says. "Physically and intellectually, but also emotionally." Cohen is a psychiatrist in the Los Angeles area. He's a tall, earnest fellow with sympathetic eyes and a quiet voice that rises in passion as he speaks of his work. "We each have a different pattern of emotional reactivity. We all have the same emotional toolbox, but the tools within it may work a bit differently, depending on the person; as with all psychological traits, there are individual differences. And sometimes, due to a genetic quirk, or as a result of a person's past, that toolbox does not serve well. I spend my days helping those whose patterns of emotion get in their way."

Cohen tells me of a new patient, Jim, who came for therapy because he had just found out that his wife was divorcing him. "This upset me," Jim said to Cohen, as if that would not have been obvious.

At their first session, Jim explained that this was his third wife. "We had a fine marriage," he said. "Then I came home one day, her bags were packed, and she was gone. I had no warning, no idea she'd been contemplating that."

Jim said he couldn't understand why his wife wanted a divorce. He was so out of touch with his wife's feelings that even now that she'd left, he still didn't recognize that there had been issues. But

Cohen doesn't probe. He doesn't like to interrupt his patients. He'd rather see where they go themselves.

"I loved her, and she loved me, too," Jim continued. "She never loved another man the way she loved me." And on top of all that, Jim said, they had three kids, "wonderful kids." How could leaving him possibly make any sense?

Jim spoke with great conviction. But Cohen was pretty sure that Jim was not the ideal mate he believed he had been. He finally pressed Jim, and Jim conceded that he'd had affairs.

"That was all her fault," Jim said. "She wasn't responsive to me." And then he added, "I think my wife is an alcoholic. She's the one who really has problems. I think that's the root of all this." By the end of the session, it came out that Jim's kids weren't talking to him, either. That seemed to be as much a mystery to Jim as his wife's leaving. "I was a great father," he insisted.

Cohen observes me rolling my eyes. He knows I'm thinking the man is a jerk and a liar. But it's not that simple, Cohen tells me.

"Yes, on the surface it seems that Jim isn't being truthful. But he's not lying; he's just wrong," Cohen says. "Jim's conscious mind truly believes he was a wonderful husband and father, a wonderful person. Meanwhile, in the depths of his unconscious, the opposite is true. He is a horrid human being, and utterly unlovable."

It's a classic case of denial, Cohen explains. The mind will go to great lengths to cover up something that is painful to the heart, though there is a price to pay for a lifetime of deluding oneself.

According to Cohen, Jim's mental state is not dominated by the grandiosity and bluster that is superficially apparent. What rules Jim's life is shame—a feeling of distress or humiliation that results from making a negative evaluation about yourself and is associated with a desire to hide or escape. It is one of the most harmful emotions, and for protection Jim has created a narcissistic shell, a constellation of primitive defense mechanisms that enable him to avoid becoming aware of that intolerable unconscious view of himself.

All emotions are reactions to a circumstance or situation. They arise, they guide our thinking, and then they dissipate. But

Jim is hypersusceptible to shame, so even minor events such as a bit of gentle criticism of which most of us would take little heed can elicit in Jim a strong shame reaction. As a result, he is so often in a shame state that it almost forms an undercurrent for his life, coloring everything he does.

We all have varying tendencies toward experiencing different emotions. Cohen speaks of a patient's collective set of emotional tendencies, his "emotional profile." In the academic literature a variety of related concepts have been studied, under names such as temperament, biological sensitivity to context, stress reactivity, affective style, and emotional style.

In Jim's case his emotional profile is very strongly inclined toward shame. So much so that Cohen says shame is Jim's "dominant emotional state." The idea that your emotional profile may be dominated by a single emotion or set of emotions goes back to ancient Greco-Roman physicians, who sorted people into four types. The sanguine person was said to be positive and outgoing; the melancholic, prone to fear and sadness; the choleric, irritable and prone to aggression; and the phlegmatic, slow to excite. But that classification was too simplistic. Most people don't have an emotional profile dominated by a single emotion. They have a balanced emotional life.

Your emotional profile is a description of what it takes to trigger each particular emotion, how swiftly it builds, how intense it tends to be, and how long it generally takes to dissipate. Psychologists use the terms "threshold," "latency to peak," "magnitude," and "recovery."[1] These aspects vary among individuals, and for each person they depend on the specific emotion in question, in particular whether it is a positive or negative one.

Some of us are easily hurt or embarrassed but perhaps hard to frighten. Others may require a more extreme situation to elicit hurt or embarrassment but may be easily frightened. Some might be insulted if you tell them they look tired or their outfit looks awful; others will shrug it off. With regard to each emotion, we all have a different threshold for that response.

Latency to peak refers to the time development of an emotional reaction. Some of us are quick to become anxious, while in others anxiety builds slowly. The magnitude of emotional reactions also varies a great deal. Being cut off in traffic or cut in front of in a line at the grocery store will make some people mildly angry, and others furious. And finally, recovery describes a person's return to the baseline state. Some people can let go of certain emotions quickly, while others hold on to them. "Recovery" can be a confusing term when applied to positive emotions; it sounds odd, for example, to dub the time it takes for the good feeling you get from hearing a compliment to fade "recovery time," but that is what psychologists call it.

Together, your tendencies toward emotional response, for each emotion, form a kind of emotional fingerprint. That's your emotional profile. How does that profile develop, how can you become aware of it, and how can you alter it, if desired?

NATURE VERSUS NURTURE

When I was in college, I once brought a girlfriend over to meet my parents. She asked my mother what I was like as a child. I must have been cute, my girlfriend said. To this my mother answered, in her thick Polish accent, "Cute? Yeah, but oy, you think Lenny gets in trouble *now*? You should have seen him then! He's a good boy, don't get me wrong. But when he was three, he tried to shave with my husband's razor. He cut up his face, and that was just the beginning. So many times we take him to emergency room. Or pick him up from principal's office. He'll never change. Good you put up with it! A lot of girls, they wouldn't. I thought he would never find a girl!" With that I ushered my friend away to do something more pleasant, like shoveling snow.

My mother believed she always knew how we kids would turn out. My older brother was shy and anxious from the start, she said. My younger brother has always been friendly and chatty. And I was created curious, not in a cute way, but in the dangerous manner that killed the cat. And indeed, my older brother grew up to be a loner and my younger brother to become a doctor who was regularly scolded by the powers that be for spending too much time with his patients. I suppose I fit my mother's theory, too, the toddler who experiments with razors and then grows up to become a scientist.

Certainly, as my mother seemed to believe, there are aspects of a person's emotional profile that are innate. Infants just two to three months old smile and laugh, and exhibit frustration and anger, and different babies show different reactivities in each of those dimensions.[2] But just as certainly, our experiences growing up contribute to our developing emotional profile.

In Jim's case, his susceptibility to shame was the result of a mother who continually criticized him. When, as a toddler, he bit too hard on his mother's nipple, his mother would shriek, throw him back in his crib, and walk away in a huff. Flash forward to when Jim was shopping for a prom tuxedo. His mother found one she liked and showed it to him. When Jim didn't swoon over the choice, his mother turned on her heels, walked out to the car, and left, abandoning Jim to fend for himself. In the decades between those incidents, there were countless others of the same type, each imparting to him the message: you're awful.

"As an adult," Cohen says, "Jim is a prisoner of his childhood. His constant shaming was an extreme case, but it's not uncommon for a person's propensity to feel shame to have been shaped by such childhood incidents. The same applies to other emotions. Our emotional profile is the result of childhood experience interacting with our genetic makeup."

Though they argue about which factor dominates, psychologists have always accepted that both nature and nurture are important to our emotional development. Today we have more insight than ever into that interplay, thanks to the rise of neuroscience, which makes it possible to relate our emotional characteristics to processes and networks in the brain.

One of the first studies of the nature/nurture issue—and perhaps the most insightful—was carried out in the 1990s by Michael Meaney, a scientist at McGill University in Montreal. A pioneer in what is now called "epigenetics," Meaney discovered a mechanism through which nurture can exact its effect in a manner analogous to the genetic route that nature takes.[3]

The key to epigenetics is the fact that although genetic traits are encoded in an organism's DNA, for them to appear in the organism, the relevant section of DNA has to be activated. Scientists used to think that was automatic, but we now know that sections of DNA can be turned on or off and that this is often determined by our environment or our experiences. We may be stuck with our genes for life, but we are not stuck with the effect of those genes. They can often be modified. Epigenetics is the study of the process through which our environment and our experiences can alter the effects of our DNA.

Meaney was led to his research through a chance meeting with another McGill scientist, Moshe Szyf, at an international scientific meeting the two of them were attending in Madrid. Szyf was an expert on the manner in which chemical alterations to DNA affect gene activity. Though at the same university, they had never met, so they went to a bar for some beer. "A lot of it," Szyf said.[4]

As they guzzled their beers, Meaney told Szyf about experiments he had conducted on rats, which had revealed that rats raised by inattentive mothers tend to be more anxious than rats raised by more nurturing mothers. He also described how the activity of stress-related genes was altered in the rats who hadn't gotten much nurturing. That's when a lightbulb flashed in Szyf's mind: Could the anxiety difference between the inattentively mothered rats and the well-nurtured rats be due to epigenetics? The idea contradicted conventional thinking in both epigenetics and neuroscience. At that time, scientists who studied epigenetics believed the process was restricted to changes effected at the embryonic stage, or in cancer cells. Most neuroscientists, meanwhile, held that long-term changes in behavior result from physical changes in neural circuits and had nothing to do with DNA expression. But Meaney was intrigued by Szyf's hunch and began to pursue it, eventually also collaborating with Szyf.

BEHAVIORAL EPIGENETICS

The rats Meaney had described to Szyf had a baseline of high anxiety. They were hypersensitive to threats in their environment, and even to things that were unfamiliar or events that were unexpected. They would freeze if dropped into unfamiliar surroundings, and they'd jump a foot into the air if you startled them. When those rats had stressful experiences, they released large amounts of hormones called glucocorticoids that make the heart pump faster and prime the muscles for fight or flight. Females of that type, due to their constant state of stress, did not tend to their babies very well and did not give them the usual amount of attention.

Other rats in Meaney's laboratory were at the other end of the anxiety spectrum. Placed in a new environment, these rats would explore. Even when given an electric shock, they released only a small amount of stress hormone. The females in this group were very attentive to their babies.

Meaney had noticed that the mothers of the mellow rats had spent a lot of time licking and grooming their babies, which meant that the mothers were also of the mellow type. Meanwhile, the mothers of the anxious baby rats rarely licked and groomed them, meaning that anxious rats were born of anxious mothers. The trait of mellowness versus anxiety seemed to be passed on genetically from generation to generation. But if Szyf was correct, there was more to it than that.

Meaney conducted a clever experiment to investigate. He switched the babies at birth so that the offspring of the anxious mothers were raised by mellow mothers and vice versa. If a mellow-versus-anxious emotional profile was inherited, the switching shouldn't make any difference. But it did. The babies grew to have the trait of the mother who raised them, not the one from whose womb they'd emerged. Meaney knew from other studies that genetics were a factor, but his experiment seemed to indicate it was the mother's behavior and not the rats' genes that governed their emotional profile. What was going on?

Through physiological investigation, Meaney and his col-

leagues found that the gene governing the stress hormone receptors—a sort of "mellowness" gene—in the rat brain is altered by
a rat's early life experiences. The mother rats' licking and grooming allow that mellowness gene to be activated. If the mother is
neglectful, clusters of atoms called methyl groups attach to the
DNA segment containing the mellowness gene, and its action is
suppressed, leaving the rat susceptible to anxiety.

The work of Meaney and his colleagues provided the missing
link in the nature-versus-nurture argument, showing how nature
and nurture interact, how your experiences alter the action of
your DNA. But the idea that our experiences can alter the expression of our genes was revolutionary at the time. Many scientists
accepted it but argued that what happens in rats may not happen
in humans. So Meaney conducted another study.

He and his colleagues obtained samples of brain tissue and
extensive psychological and medical histories from people who
had suffered child abuse and subsequently taken their own lives.
He found that, as with the rats, compared with the brain tissue of
those who hadn't been abused, the abused brains contained significantly more methylation on the gene responsible for the stress
hormone receptor. As with rats, stressful childhood experiences
leave adult humans less able to cope with adversity (and thus more
vulnerable to suicide). Meaney had discovered that one's emotional profile arises from genetic predisposition plus epigenetics,
which seems to be an important mechanism through which nurture exerts its effect.

The work of Meaney and his colleagues led to a new field
called behavioral epigenetics. It provides great hope for those suffering from emotional issues, showing that even if a tendency is
inherited, it can be possible to change one's brain to overcome it.

Your emotional profile may be most influenced by early childhood experience. In adulthood, your emotional profile doesn't
normally change much. You head into the adult world more
or less set in your ways unless you work hard to alter that. But
Meaney's work showed that you *can* alter it. The emotional profile
you inherit from your younger self doesn't have to be permanent.

You can transform your brain. The first step in that process is to find out what your emotional profile is.

YOUR EMOTIONAL PROFILE

Emotion usually affects our thoughts, decisions, and behavior for the good; to be immune to any set of emotions is not a good thing. On the other hand, to feel too strongly can complicate life. There are no rights and wrongs in an emotional profile, but there are some profiles that can make life easier and others that can lead to unnecessary pain or interfere with the life you'd like to have. In the next chapter we'll examine how you can regulate your emotions and influence the emotions of others. But first it's interesting and useful to learn about your profile. Some of you are reading this book in order to apply its lessons to your life; others are simply interested in a better understanding of human nature. Even if you are in the latter group, understanding your own profile is useful because to know yourself in this way will help you understand others.

Both clinicians and emotion researchers emphasize that one of the most notable aspects of human emotion is how much it varies across individuals. There is an enormous range of emotional profiles; different people react very differently to similar circumstances and challenges. In order to study those individual differences, over the years psychologists and psychiatrists have developed and published, in academic journals, a number of "inventories." These are questionnaires that can be used to characterize your emotional profile along many different dimensions. They weren't developed as part of a systematic exploration of emotion; rather, each questionnaire was developed by individual researchers seeking to understand the particular emotion in which they specialized. The seven I've included here have proved to be the most influential, as judged by their popularity in the academic literature. They measure shame, guilt, anxiety, anger, aggression, joy/happiness, and romantic love/attachment. Filling out the questionnaires will help you identify the manner in which you tend to respond to various emotionally charged circumstances in everyday life.

These inventories were not the work of self-help-book authors. They were created by scientists interested in understanding the core human psyche. Some were designed for use in studying those with physical or psychological disorders that disrupt their emotional lives, but even then the inventories had to first be validated by studying the responses of those without such disorders. The researchers developed the questionnaires through extensive trial and error and typically validated them in studies of hundreds or thousands of subjects, and in one case more than ten thousand. Through those studies, psychologists were able to confirm the consistency and stability of subjects' scores. By consistency here I mean that if you take the questionnaire on Tuesday and again on Thursday, you ought to get a similar score; by stability I mean that if you take the test today and six months from now, you ought to also get a similar result (unless a major life-changing event, or therapy, has intervened).

In what follows, I'll present those research-validated assessment tools. You may not wish to answer all the questionnaires, or you may wish to spread them out and answer them over a period of time as you continue on to the next chapter. They don't require you to have a lot of insight, but they do require you to be honest about your past behavior and feelings. If you answer them honestly, you should learn a lot about yourself.

Many people are surprisingly ignorant of their emotional profile. If your results don't ring true, you could take them with a grain of salt. But be willing to at least consider the possibility that they are accurate, for results that surprise you might be opening your eyes to tendencies you weren't aware of.

It's also possible to use the questionnaires to get insight into others, if you are close enough to them to assess how they would answer the various questions. Conversely, it can be fun and enlightening to ask a significant other or someone else who is close to you to try to answer the questions from your point of view, and then compare the result with your own. That can also serve as a reality check on the honesty of your own answers.

The formats and scoring of the questionnaires are all similar, but not identical. That's because each was created by a different group of researchers, each of whom had their own approach. So you will need to read the instructions and statements carefully. Note, in particular, that some of the statements in the questionnaires are phrased positively and others negatively.

Don't take too long on individual questions: there are no right or wrong answers, and no trick questions. Just give the answer that is true for you in general or most of the time. And, though you can feel free to skip whichever questionnaires you don't want to take, please don't skip any of the items in a questionnaire that you do take. That would invalidate the scoring.

Some of the questions may begin by positing something that you wouldn't do. For example, one says, "You break something at work and then hide it," and then asks you to rate the likelihood you'd respond in a certain way to having done that. In such cases, do your best to rate what you would do even though you don't feel you'd ever be in that situation. The questionnaire is about your responses, even if the situation you are responding to is unlikely to ever occur.

You may also have trouble making up your mind in response to some items: Is what is described something I would "2 = somewhat likely feel" or "3 = very likely feel"? Such difficulties are normal, and you can just pick one or the other. The idea is that there are enough questions that such ambiguities will cancel each other, and in any case the inventories are not that precise; a difference of a point up or down isn't significant. So don't overthink it: the first answer that comes to mind is probably the appropriate one for you. And, finally, it's important to realize that the questionnaires are designed to measure propensities or capacities, not behaviors or your current, fleeting emotional feelings.

THE SHAME AND GUILT INVENTORIES[5]

Below are eleven situations that might occur in everyday life. Try to imagine yourself in the situation described. Each is followed by a description of two common ways (*a* and *b*) in which a person might respond. You are not being asked to choose between them. Rather, you are to rate how likely it is that you would respond as described by both. It is possible (and acceptable) that both *a* and *b* are ways you'd be likely to respond, in which case you would rate them both 5, or that neither is, in which case you would rate them both 1.

Directions: Rate each response with a number from 1 to 5, where

1 = very unlikely 2 = somewhat unlikely 3 = sometimes
4 = somewhat likely 5 = very likely

1. You make plans to meet a friend for lunch. At five o'clock, you realize you stood him up.
     a) You would think: "I'm inconsiderate." _____
     b) You would think you should make it up to your friend as soon as possible. _____
2. You break something at work and then hide it.
     a) You would think about quitting. _____
     b) You would think: "This is making me anxious. I need to either fix it or get someone else to." _____
3. At work, you wait until the last minute to plan a project, and it turns out badly.
     a) You would feel incompetent. _____
     b) You would feel: "I deserve to be reprimanded for mismanaging the project." _____
4. You make a mistake at work and find out a co-worker is blamed for the error.
     a) You would keep quiet and avoid the co-worker. _____
     b) You would feel unhappy and eager to correct the situation. _____
5. While playing around, you throw a ball and it hits your friend in the face.

a) You would feel inadequate that you can't even throw a ball. _____

b) You would apologize and make sure your friend feels better. _____

6. You are driving down the road, and you hit a small animal.
    a) You would think: "I'm terrible." _____
    b) You'd feel bad you hadn't been more alert driving down the road. _____

7. You walk out of an exam thinking you did extremely well. Then you find out you did poorly.
    a) You would feel stupid. _____
    b) You would think: "I should have studied harder." _____

8. While out with a group of friends, you make fun of a friend who's not there.
    a) You would feel small . . . like a rat. _____
    b) You would apologize and talk about that person's good points. _____

9. You make a big mistake on an important project at work. People were depending on you, and your boss criticizes you.
    a) You would feel as if you wanted to hide. _____
    b) You would think: "I should have recognized the problem and done a better job." _____

10. You are taking care of your friend's dog while they are on vacation and the dog runs away.
    a) You would think, "I am irresponsible and incompetent." _____
    b) You would vow to be more careful next time. _____

11. You attend your co-worker's housewarming party and you spill red wine on their new cream-colored carpet, but you think no one notices.
    a) You would wish you were anywhere but the party. _____
    b) You would stay late to help clean up the stain. _____

TOTAL for all the (a) responses = SHAME SCORE _____

TOTAL for all the (b) responses = GUILT SCORE _____

The range of scores is 11–55 for both guilt and shame; over a number of studies, about half the respondents scored in the range of 25–33 on the shame scale, and 42–50 on the guilt scale; women have typically scored a couple points higher than those averages on both tests, and men a couple points lower.[6]

Until recently there hasn't been much systematic empirical research into shame and guilt. This inventory, created by the field's leading researchers, was made to help remedy that. Shame and guilt are two emotions closely linked to the self in relationship with others.[7] Shame represents concern about how you and others see you, while guilt has to do with concern about how your action affected others. Shame, as I mentioned earlier, is associated with a desire to hide or escape, while guilt correlates with the wish to apologize or repair. The role of both in our social interaction is to inhibit misdeeds and wrongdoing and to foster repair, apology, and atonement. In an interesting longitudinal study, for example, it was found that guilt-prone fifth graders were less likely than the average child to drive under the influence of alcohol later in life and more likely to do community service.

The tendency toward both guilt and shame has its roots in our earliest family experiences; it seems to be passed down from our parents, and especially our fathers. Shame emerges at about age two, but guilt requires stronger cognitive abilities and is not typically seen before age eight. Shame is a painful feeling that usually has a negative impact on relationships with others. People prone to shame can tend to blame others for negative events, and may also be prone to anger and hostility, and generally less able to empathize with others. Guilt-prone individuals, on the other hand, tend to be less prone to anger, more likely to express their anger in a direct way, and appear better able to empathize with others. They also tend to accept responsibility for negative outcomes.

THE ANXIETY INVENTORY[8]

Directions: Write the number that best applies after each item.

1 = almost never 2 = sometimes 3 = often 4 = almost always

1. I feel secure. ____
2. I am calm, cool, and collected. ____
3. I make decisions easily. ____
4. I am content. ____
5. I am happy. ____
6. I feel satisfied with myself. ____
7. I am a steady person. ____
8. I feel pleasant. ____
9. I feel rested. ____
10. I worry too much over something that really doesn't matter. ____
11. I feel nervous and restless. ____
12. I get in a state of tension or turmoil as I think over my recent concerns and interests. ____
13. I lack self-confidence. ____
14. I feel that difficulties are piling up so that I cannot overcome them. ____
15. I feel like a failure. ____
16. I take disappointments so keenly that I can't put them out of my mind. ____
17. I wish I could be as happy as others seem to be. ____
18. Some unimportant thoughts run through my mind and bother me. ____
19. I have disturbing thoughts. ____
20. I feel inadequate. ____

The anxiety inventory is structured so that agreement in questions 1–9 indicates low anxiety, while for questions 10–20 agreement suggests that anxiety is present. As a result, the scoring is slightly more complicated than in the shame and guilt inventories. Here is how it works:

1) Add your response for questions 1 through 9 to get: ____
2) Subtract the total in line 1) from 45 to get: ____
3) Add your response for questions 10 through 20 to get: ____
4) Add the results from line 2) and line 3) to get your
ANXIETY SCORE = ____

The scores on the anxiety inventory range from 20 to 80; the average is 35; about half of all subjects score between 31 and 39.[9] Depressed patients, who often suffer from anxiety, typically score in the 40s or 50s.[10]

Anxiety is evoked by the perception of a threat. In contrast to fear, which is a response to a specific, identifiable, and imminent danger, anxiety comes from the perception of potential and unpredictable threat, one that may pose a low probability of actual harm, or be vague or ambiguous, or not have a clear source. It is more common, therefore, that people live in a state of chronic anxiety than chronic fear. From an evolutionary perspective, both emotions help shield us from harm, but in very different ways. Fear stimulates a defensive reaction—the fight-or-flight response—and it quickly subsides as the threat disappears. Anxiety is associated with less direct coping methods and can persist for some time. It protects you by promoting the anticipation of, and preparation for, a potentially harmful situation.

Having too great a tendency toward anxiety is unhealthy because it causes stress, and a chronic oversupply of stress hormones causes a wide variety of medical problems. But while a tendency toward high anxiety leads to an increased mortality risk, those with abnormally low anxiety also have higher death rates: low anxiety reduces the likelihood that individuals will seek help in the presence of a threat, or take prudent action to avoid it. They may not rush to the doctor when finding that lump under the skin or may be more likely to smoke or engage in other risky behavior.

THE ANGER AND AGGRESSION INVENTORIES

Just as shame and guilt are linked and need to be considered together, anger and aggression are related, for aggression is a response to anger. The anger and aggression inventories, below, are linked in the emotional profile and should be looked at as a pair.

Directions: Write the number that best applies after each item.

1 = extremely uncharacteristic of me
2 = somewhat uncharacteristic of me
3 = my answer could go either way
4 = somewhat characteristic of me
5 = extremely characteristic of me

1. I flare up quickly but get over it quickly. ____
2. When frustrated, I let my irritation show. ____
3. I sometimes feel like a powder keg ready to explode. ____
4. I am NOT an even-tempered person. ____
5. Some of my friends think I'm a hothead. ____
6. Sometimes I fly off the handle for no good reason. ____
7. I have trouble controlling my temper. ____

TOTAL ANGER SCORE = ____

1. I tell my friends openly when I disagree with them. ____
2. I often find myself disagreeing with people. ____
3. When people annoy me, I may tell them what I think of them. ____
4. I can't help getting into arguments when people disagree with me. ____
5. My friends say I'm somewhat argumentative. ____

TOTAL AGGRESSION SCORE = ____

Scores on the anger inventory range from 7 to 35. The average score is 17; about half of all people score between 13 and 21.

The scores on the aggression inventory range from 5 to 25; the average score is 15, and about half of all subjects score between 12 and 18.[11]

People tend to view anger and aggression as destructive or at least counterproductive. But from the evolutionary point of view, they could only have evolved if they enhanced the chances a person would survive and reproduce. It is useful to understand that evolutionary origin in order to understand anger and aggression in ourselves and others.

Across the animal world, it is the access to resources such as food, water, and mates that determines which animals survive and pass on their genes. Though in the modern human world that access is not usually decided through the threat of force, when we were evolving, and in most animal societies, it was the threat of force that determined who got what. Perhaps the most crucial function of anger and aggression, in our evolution, was to establish access to those necessary resources for an individual and his or her progeny.

Anger spurs someone to take action when access to survival needs is challenged, or when another individual blocks the attainment of a goal. An angry response often seems disproportionate to its trigger, but from the point of view of survival that makes sense because an angry reprisal is designed to deter not just the current threat but the sum of all similar threats that would occur in the future if the angry actions were not taken.

Aggression is an important defense reaction that can be activated in a number of different contexts, for example, when an individual threatens a mother's young. The type of aggression tested in the above inventory is verbal aggression, a modern form that might not have existed tens of thousands of years ago but is certainly relevant in today's society. A low score may indicate that you are hesitant to assert yourself. A score that's very high is a sign that you may be considered argumentative by others.

Neither anger nor aggression will necessarily have the same effect today that it did in our ancestral environment, and they both can career out of control. If you scored high on the anger or aggression scales, or if you experience a lot of life stress, which lowers your threshold for what will provoke those emotions, it is important to be vigilant about emotional regulation. Not only might you do something that you'll regret, but also you may suffer from various physical issues such as migraines, irritable bowel syndrome, and high blood pressure, which are caused by excess anger. In fact, studies show that those who habitually react with anger or aggression are significantly more likely than calm people to have an early heart attack.

I will talk about general methods for emotion control in the next chapter, but with regard to these emotions in particular there are two specific approaches that are quite effective. One is to remove yourself from the situation, take a break, take a walk, take several deep breaths, let the passage of time calm you. The other is to have compassion for the subject of your anger. Say someone approaches you with a weapon and demands money. You could get angry and vilify the person, or you could focus on the unhappiness and hardship that might have driven the person to that extreme. That's what the NBA player Lou Williams did when a mugger tapped on the window with a gun demanding money while Williams's car was stopped at a red light in North Philadelphia. Williams engaged the man, who eventually told him, "I just got out of jail. I'm hurting. I'm hungry. All I got is this gun." In the end, the mugger backed off, and Williams took him to dinner. The Dalai Lama, among others, advocates that approach. Once, a woman on her way to see him saw a man on the street beating his dog.[12] She asked the Dalai Lama about it when she met with him. He said, "Compassion means being sorry for the man as well as the dog." By defusing anger, that compassion benefits all involved.

The Oxford Happiness Questionnaire[13]

Directions: Below are a number of statements about happiness. Please indicate how much you agree or disagree with each by entering a number according to the following code:

1 = strongly disagree 2 = moderately disagree 3 = slightly disagree
4 = slightly agree 5 = moderately agree 6 = strongly agree

1. I don't feel particularly pleased with the way I am. ____
2. I rarely wake up feeling rested. ____
3. I am not particularly optimistic about the future. ____
4. I do not think that the world is a good place. ____
5. I don't think I look attractive. ____
6. There is a gap between what I would like to do and what I have done. ____
7. I feel that I am not especially in control of my life. ____
8. I do not find it easy to make decisions. ____
9. I do not have a particular sense of the meaning and purpose in my life. ____
10. I do not have fun with other people. ____
11. I don't feel particularly healthy. ____
12. I do not have particularly happy memories of the past. ____
13. I am intensely interested in other people. ____
14. I feel that life is very rewarding. ____
15. I have very warm feelings toward almost everyone. ____
16. I find most things amusing. ____
17. I am always committed and involved. ____
18. Life is good. ____
19. I laugh a lot. ____
20. I am well satisfied about everything in my life. ____
21. I am very happy. ____
22. I find beauty in some things. ____
23. I always have a cheerful effect on others. ____
24. I can fit in everything I want to. ____

25. I feel able to take anything on. ____
26. I feel fully mentally alert. ____
27. I often experience joy and elation. ____
28. I feel I have a great deal of energy. ____
29. I usually have a good influence on events. ____

This inventory is structured so that agreement in questions 1–12 indicates low happiness, while for questions 13–29 agreement suggests happiness. As a result, the scoring is slightly more complicated than in most other inventories. Here is how it works:

1) Add your response for questions 1 through 12 to get: ____
2) Subtract the total in line 1) from 72 to get: ____
3) Add your response for questions 13 through 29 to get: ____
4) Add line 2) and line 3) to get your
   TOTAL HAPPINESS SCORE: ____

The score on the Oxford Happiness Questionnaire can range from 29 to 174. The average score is about 115; most scores fall between 95 and 135.[14]

The happiness questionnaire provides a measure of your baseline happiness level. That's a set point that comes from your DNA. It determines only your "susceptibility" to being happy. Whether you *are* happy, and how happy you are, will depend not just on your baseline but on other factors, such as your external circumstance and your behavior.

People tend to overrate the importance of external circumstance, of life events, in making us happy. We assume that making more money, driving a nicer car, or having our favorite sports team win the world championship will create much more happiness than they really do. Likewise, we assume that losing a job, breaking up with a romantic partner, or having our team lose a major contest will make us more unhappy than they really do.

But research shows that while circumstances and events affect us, they don't change our happiness level as much or for as long as we expect. For example, in a classic study, researchers asked a hundred persons from the *Forbes* list of wealthiest Americans to report on their level of happiness, along with a hundred control persons, selected from telephone directories.[15] The richest Americans, earning more than $10 million per year, were only marginally happier than their average counterparts.

Research suggests that your happiness set point as well as circumstances and recent events accounts for much but not all of your happiness level. What about the rest? That's due to our behavior, and the good news is, in contrast to the others, this factor is very much under our control. It has been studied extensively by happiness researchers in recent years.[16] So if you scored lower than you'd like on the happiness inventory, or if you simply feel you'd like to be happier, here are some behaviors recommended by a leader in the field, Sonja Lyubomirsky: devote time to family and friends; focus on and express gratitude for all you have; engage regularly in acts of kindness toward others; cultivate optimism as you ponder your future; savor life's pleasures and try to live in the present moment; exercise weekly or daily; try to find and commit to lifelong goals, whether that be social activism, teaching children, writing a novel, or maintaining a wonderful garden.[17] Says Lyubomirsky, "Consider how much time and commitment many people devote to physical exercise, whether it's going to the gym, jogging, kickboxing, or yoga . . . [I]f you desire greater happiness, you need to go about it in a similar way. In other words, becoming lastingly happier demands making some permanent changes that require effort and commitment every day of your life."

THE ROMANTIC LOVE/ATTACHMENT INVENTORY

This inventory measures the extent to which you are "susceptible" to love and attachment—comfortable being close to others and in an intimate, loving relationship. If you are in a romantic relationship, try to answer each question in general, and not according to the specifics of your current relationship.

Directions: Rate the items below according to the following scale:

1 = strongly disagree . . . 2 . . . 3 . . . 4 . . . 5 . . . 6 . . . 7 = strongly agree

1. I feel comfortable sharing my private thoughts and feelings with my partner. ____
2. I am very comfortable being close to partners. ____
3. I find it relatively easy to get close to my partner. ____
4. It's not difficult for me to get close to my partner. ____
5. I usually discuss my problems and concerns with my partner. ____
6. It helps to turn to my partner in times of need. ____
7. I tell my partner just about everything. ____
8. I talk things over with my partner. ____
9. I feel comfortable depending on partners. ____
10. I find it easy to depend on partners. ____
11. It's easy for me to be affectionate with my partner. ____
12. My partner really understands me and my needs. ____
13. I prefer not to show a partner how I feel deep down. ____
14. I find it difficult to allow myself to depend on partners. ____
15. I don't feel comfortable opening up to partners. ____
16. I prefer not to be too close to partners. ____
17. I get uncomfortable when a partner wants to be very close. ____
18. I am nervous when partners get too close to me. ____

In this inventory, agreement with items 1–12 represents attachment, while agreement with items 13–18 indicates attachment avoidance. The scoring is as follows:

1) Add your response for questions 1 through 12 to get: ____
2) Add your response for questions 13 through 18 to get: ____
3) Subtract the total in line 2) from 48 to get: ____
4) Add line 1) and line 3) to get your
   TOTAL LOVE/ATTACHMENT SCORE: ____

> Scores on the Romantic Love/Attachment Inventory
> range from 18 to 126. The average score is 91.5. About
> half of all scores fall between 78 and 106, so if you fall
> below that, you are less open to intimate attachment than
> most, and if you're above it, you are more open to it than
> most.[18]

The emotional state of love has an enormous effect on brain chemistry.[19] As you might expect, to just see a loved one causes your brain to release dopamine—activating the wanting apparatus in your reward system. But love in the brain is also distinguished by what it deactivates. One such set of regions that is deactivated is associated with negative emotion, providing that feeling of being on cloud nine. Another is a region associated with social judgment, making those in love generally less critical of others. And another has to do with the ability to distinguish between the self and others, enabling the feeling that you and your beloved are as one. And so your state of mind when deep in love will be unusually biased toward the welfare of your beloved, in comparison to your own. Why did nature endow us with this complex and life-altering mental state? How does this contribute to human survival and reproductive success?

Anthropologists tell us that romantic love is a very old emotion. It is said to have evolved about 1.8 million years ago. Mammalian reproduction requires a particularly intense investment of maternal time and energy and a commitment to a specific infant. Being attached to a mate increases not just the couple's own ability to survive but also that of their offspring. The women were more able to attend to their children's survival, while the men assisted women with food gathering, shelter procurement, protection, and the imparting of those skills to their offspring.

Today love seems to vary little across the globe. In a survey, anthropologists found evidence of romantic love in 147 widely varying cultures.[20] Even among the Hadza, an isolated tribe of pretechnological hunter-gatherers in Tanzania, there is love, marriage, and commitment. What's more, evolutionary psychologists who studied the Hadza found that the degree of commitment among partners correlates with the number of surviving children—that is, their "reproductive success."[21] Or as the "notoriously grumpy" poet Philip Larkin put it, "What will survive of us is love."[22]

YOUR EMOTIONAL PROFILE

Now that you've assessed your tendencies, you can review your scores and think about your emotional profile. You might be pleasantly high on joy and love, but find you have tendencies toward shame and guilt. Or you may discover (or confirm) that you are unusually anxious.

There is no right and wrong in these scores. Everyone is different, and those differences are part of who we are. Certainly we need not aim to be in the middle of the pack in each aspect of our profile. I have friends who are chronically anxious and proud of it; they claim it helps them be more cautious and avoid trouble. I know others who are profoundly joyful and optimistic, which often leads them to suboptimal decisions but is nonetheless a happy state of being. Others who have taken these tests have found them eye-opening, helping them to become more aware of their feelings and the reasons behind some of their actions. And once aware, they sometimes take action to try to change some of the aspects that are standing in the way of a fuller, richer life.

Your emotional profile is the result of a complex interplay of nature and nurture, of the physical makeup of your brain and the experiences that have impacted that. We all react to our emotional state but also have an ability to control it. That control, or regulation, can be both conscious and unconscious. What's more, processes that are initially under our control and voluntary can evolve to become more automatic with practice. Whatever your profile, knowing where you stand is the first step toward understanding how your emotions affect your life, and deciding whether you'd like to work on changing that, which is the topic of our final chapter.

9

Managing Emotions

In October 2011 a cheerleader at Le Roy High School in western New York woke from a nap to find her face twitching and her chin jutting forward in uncontrollable spasms.[1] A few weeks later she was still experiencing those symptoms when her best friend, also a senior and a cheerleader, began to stutter after waking from a nap. A little later, she, too, began to twitch. Her arms flailed. Her head jerked back and forth. Two weeks after that a third case erupted. Soon there were a dozen teenage girls afflicted with the malady.

The nature of the symptoms raised the possibility of a neural illness or toxic contamination. One neurologist suspected it was caused by a rare type of immune response to a streptococcal infection. Others suspected something in the water at the school or in the soil on school grounds. Or possible leakage from a forty-year-old cyanide dump in the neighborhood. Investigators searched the academic literature for past episodes of contagious neurological tics. The New York State Department of Health investigated. So did Erin Brockovich, famous, despite her lack of formal education, for winning a $333 million settlement from Pacific Gas and Electric for environmental contamination. Over a period of months, researchers scrutinized family medical histories, past illnesses, and possible exposure to toxic substances. Drinking water from the school building was tested for fifty-eight organic chemicals, sixty-three pesticides and herbicides, and eleven metals. The indoor air quality was scrutinized, and mold was searched for.

The medical detectives found nothing out of the ordinary and were left with several nagging questions. Why was the sickness appearing almost exclusively in adolescent girls? Why weren't their parents or siblings affected? And why would the symptoms develop suddenly when the toxins, if any, had been around for years or decades? In the end, most experts agreed that the girls suffered from a kind of psychological contagion.

Though it doesn't often make the news, outbreaks of what is technically labeled mass psychogenic illness are more common than one might think. For example, in 2002 ten girls at a high school in North Carolina had similar symptoms. So did nine girls at a high school in Virginia in 2007. But the phenomenon is not restricted to any age-group, or gender, or any particular culture. Analogous episodes have been observed worldwide, even among hunter-gatherer tribesmen in New Guinea.[2] The syndrome can arise in any group whose members have some social connection and are experiencing long-term or intense anxiety.

A much milder and more everyday kind of contagion is that described by Adam Smith in 1759: "When we see a stroke aimed, [at] the leg or arm of another person, we naturally shrink and draw back . . . our leg or our own arm."[3] Smith felt that such imitation was "almost a reflex." He was right. We are hardwired to feel what others do. In fact, brain imaging studies show that the brain structures activated when we feel our own emotions are automatically activated when we observe those emotions in others.[4]

The spread of emotion from person to person or throughout an organization or even an entire society is an important subfield in the new science of emotion, generating in recent years a tenfold increase in the number of annual studies. Psychologists call the phenomenon "emotional contagion."

You chat with a colleague. You notice you feel a bit uncomfortable. You're becoming anxious. As you walk away, you remember you were feeling fine before you started the chat. You realize that your colleague often has that effect on you. She tends toward anxiety, and after speaking with her, so do you. Why does that happen?

Historically, human survival depended on the ability to function within a social context. We have to understand others and find ways of forming a connection. Synchronizing emotions helps facilitate that connection. As a result, humans, like other primates, are natural mimics. Partners in conversation tend to sync rhythms. When babies open their mouths, mothers tend to open theirs, too.[5] People imitate smiles, expressions of pain, affection, embarrassment, discomfort, and disgust. Even laughter is contagious. That's why television comedies have laugh tracks and late-night talk-show hosts do their monologues in front of studio audiences that have been primed (and implored) to laugh. To those listening at home, the same jokes that seem hilarious with an audience responding in the background often fall flat when there is none.

The kind of mimicry I'm talking about arises not from conscious intention but from our unconscious. We aren't aware we are doing it. Some of the mimicking even depends on reaction times that are not achievable through conscious intent. For example, in a classic study, it took Muhammad Ali 190 milliseconds to spot a signal light and 40 milliseconds more to throw a punch in response.[6] But studies of college students engaged in social interaction show that they sometimes synchronize their facial and body movements to those of others within 21 milliseconds. That lightning synchrony is only possible because it comes from subcortical brain structures that are outside our conscious control. In fact, people who consciously try to mirror others usually look phony.

One of the effects of emotional contagion is that people's degree of happiness tends to reflect that of their friends, family, and neighbors. We are, in a sense, whom we hang out with. At least that's the conclusion of a recent collaboration between Harvard and the University of California, San Diego that followed the lives of 4,739 individuals over a period of twenty years.[7] The subjects in that study were not a random group of strangers; they were a huge social network. For each subject, the group included, on average, 10.4 others to whom that individual had some social tie—family members, neighbors, friends, even friends of friends—for a total of more than 53,000 interconnections. The subjects were interviewed every two to four years to ascertain their degree of happiness and document any changes in their social ties. The data was computerized and analyzed employing the sophisticated mathematics of network analysis. The conclusion: people surrounded by happy people tend to be happy themselves and are more likely to be happy in the future—due to the *spread of happiness,* not just a tendency for people to associate with similar individuals.

The most surprising of all the new research on emotional contagion comes from studies showing just how easily it happens. You don't have to come in contact with another person, or even talk to her over the phone; our emotions can be influenced via text alone or by social media. Consider a controversial 2012 emotion manipulation experiment Facebook conducted on its users without their knowledge. In that study, the social media company manipulated what 689,000 people saw by filtering positive or negative emotional content from their news feeds.[8] The researchers reported that when positive expressions were reduced in their news feeds, users themselves produced fewer positive posts and more negative posts. And when negative expressions were reduced, the opposite pattern occurred. A related study on Twitter (in which no one's content was manipulated) also found that people viewing negative content produced more negative posts and those viewing more positive content produced more positive posts.[9]

Like many aspects of emotion, emotional contagion, while having advantages in our evolutionary past, is sometimes suboptimal in today's society. But in one respect it delivers a very important and optimistic lesson: if other people's frowns or text messages can alter our emotional state, then we ourselves should be able to do it. And research shows that we can indeed take control of our emotions.

MIND OVER EMOTION

Our emotions take us to the depths of sorrow and the heights of joy. They are the dominant driver behind our choices and behaviors, the reason we formulate and achieve our goals. But they can also be the number one thing that derails us. It's okay to feel heart-wrenching sadness when reminded of the loss of a loved one. But not okay to feel that because you can't open a jar of marinara. One of the recurring themes in the study of emotion, and in this book, is that emotions are a necessary part of our existence and usually beneficial, but not always. Because for the most part they evolved in an era when our lives were much different, there are bound to be times when they may not be optimal for our needs today. In particular, overly intense emotional states can have a downside. Anxiety developed to make us more careful, but it can also trigger panic. Sadness at a loss reminds us of what is important, but it can crowd out any thoughts of hope or optimism and turn into depression. Anger can motivate you to address the situation that angered you, and it can increase adrenaline levels to give you a burst of strength, but it can also cause you to act in a manner that alienates others, which may frustrate your goals.

We all run into situations in which modulating emotion would be beneficial. There will be times when it would be best to hide or suppress our feelings because they will be seen by others as unprofessional or inappropriate, when for the sake of our own well-being we'll want to lower the intensity of what we are feeling. Studies of emotional intelligence show that the most successful business, political, and religious leaders are usually the ones who can control their emotions and use them as tools when they interact with others. While IQ scores may correlate to cognitive abilities, control over and knowledge of one's emotional state is what is most important for professional and personal success.

The ability to regulate emotions is a particularly human trait. Even simple animals employ many of the same neurotransmitters that we do, and emotion in many higher animals is associated with brain circuitry analogous to that in humans. Anxious mice calm down when you give them Valium; octopuses become amorous when given Ecstasy; and psychotropic drugs that work on humans often have the same effect on rats. But these animals don't have the ability to effect such changes themselves. Nor can they modulate, delay, or hide whatever it is they are feeling. Most animals react immediately and without disguise to whatever emotion arises in them. While humans can moderate, enhance, fake, or suppress emotion, a cat won't pretend it likes its food when it doesn't, nor will it suppress its feelings if you annoy it. This is one of the glaring differences between our emotion system and theirs.

In humans, emotion regulation has physical as well as psychological benefits. For example, it is correlated to better physical health, especially with regard to cardiac disease.[10] In one thirteen-year study of older men, those who had the lowest levels of emotion regulation had a 60 percent greater chance of suffering a heart attack than those who were skilled at self-regulation. Scientists aren't yet sure of the mechanism, but they suspect that emotion regulation leads to less activation of your body's stress response system. When you are in imminent physical danger, your stress response prepares you for conflict. It raises your blood pressure and heart rate, tightens your muscles, dilates your pupils so that you can see better. That's useful if you're about to face the onslaught of hyenas on the grasslands of our ancestors; it's not so useful when the attack is verbal and coming from your boss. And it comes at a cost: the response occurs through the release of stress hormones, which have an inflammatory effect that has been tied to cardiovascular disease and other illnesses.

Given the benefits of being able to manage your emotions, it's not surprising that over time people have employed many methods for achieving that end. Some work; others don't. Only in the last decade or two have research psychologists focused on sorting that out by studying and validating the efficacy of the various approaches. In what follows, I'll talk about three of the most effective: acceptance, reappraisal, and expression.

ACCEPTANCE: THE POWER OF STOICISM

Consider the story of James Stockdale. In September 1965, Stockdale was a naval wing commander on his third tour of combat duty over North Vietnam.[11] Flying just above the treetops at nearly six hundred miles per hour, his A-4 Skyhawk jet ran into a barrage of flak. The flak took out the Skyhawk's control system, so Stockdale couldn't steer. The plane caught fire. He decided to eject.

As he glided down on the short parachute ride to the village below, he realized how little control he'd have over his life for the foreseeable future. He recalls thinking, "I'm going right now from being the Wing Commander, in charge of a thousand people . . . and beneficiary of goodness knows all sorts of symbolic status and goodwill, to being an object of contempt . . . a criminal [in their eyes]."

It didn't take long for his new life to materialize. Upon landing, Stockdale was so badly beaten by a crowd that his leg was broken and he walked with a limp for the rest of his life. Knocked down, kicked, and bound with tourniquet-tight ropes, he was hauled off to a North Vietnamese prison where he was held for seven and a half years, even longer than his fellow prisoner and later friend Senator John McCain. In that period Stockdale was tortured fifteen times.

Years of torture and deprivation are bound to take an emotional toll. It's difficult not to be overcome by terror, pain, sadness, anger, anxiety. But to his fellow prisoners, Stockdale was a rock. The only wing commander to survive an ejection, he was the senior officer and became the clandestine leader of what would grow to be a prison population of nearly five hundred pilots. After the war ended, Stockdale was able to pick up the pieces, rise to the rank of vice admiral, and go on to be Ross Perot's running mate in the 1992 presidential election. How did he cope so successfully with his brutal POW state of existence?

Stockdale said that after ejecting from his jet, he figured he had about thirty seconds before he landed on the main street of that little village. And so, he later wrote, "I whispered to myself: 'Five years down there at least. I'm leaving the world of technology and entering the world of Epictetus.'"

Stockdale had studied that ancient philosopher at Stanford. There, a professor had shown him a copy of the *Enchiridion of Epictetus*, the handbook of the Greek philosophy of stoicism. It became his bible and was on his bedside table throughout the three years he spent on the aircraft carrier before being shot down.

Stoicism is often misinterpreted. It is associated with the idea that wealth or even comfort is bad. But that's not what stoicism dictates. Stoic philosophy warns us not to be overly tied to creature comforts, not to be addicted to our wealth, or anything material, but it doesn't demonize those things. Stoicism is also sometimes said to hold that one should seek to avoid emotion, but that's not quite right, either. What the Stoics believed was that you should not be psychologically enslaved to your emotions: don't be manipulated by them, be actively in command.

Epictetus wrote, "A man's master is he who is able to confirm or remove whatever that man seeks or shuns."[12] If you depend on no one except yourself to satisfy your desires, you will have no master other than yourself and you will be free. Stoic philosophy was about that—taking charge of your life, learning to work on those things that are within your power to accomplish or change and not to waste energy on things you cannot.

In particular, the Stoics warned against reacting emotionally to what is outside your control. Often, Epictetus argued, it's not our circumstances that get us down but rather the judgments we make about them. Consider anger. We don't get angry at the rain if it spoils our picnic. That would be silly because we can't do anything about the rain. But we often do get angry if someone mistreats us. We usually can't control or change that person any more than we can banish the rain, so that is equally silly.

More generally, it is just as futile to tie our feelings of well-being to altering another individual's behavior as it is to tie them to the weather. Epictetus wrote, "If it concerns anything not in our control, be prepared to say that it is nothing to you."[13] If you truly come to accept that philosophy, and to integrate it into your way of life, you can avoid or mitigate many energy-wasting emotional episodes. But you have to train your mind to embrace it—not just to know it intellectually, but to believe it in your core. If you do, you can change your emotional response system.

After he landed in the POW camp, that philosophy helped Stockdale accept his new existence. He concerned himself not with the horrors of his plight but with what he could do to survive and make his life better. He let go of his anxiety about what would happen next. He overcame his fear of torture by accepting that he could not cause it to stop, assuming it would happen again, and focusing on what he could do to survive it.

Acceptance is the heart of the stoic approach: you can lessen emotional pain if you accept that the "worst" may happen and focus only on what you can do to respond in a positive way. That allows emotion to motivate you rather than sabotaging you. The Stockdale story is a single instance, but modern researchers have studied the technique in controlled experiments, and it is powerful.

In one study students were recruited to play a simple matching game.[14] Now and then the game was interrupted, and they were given a choice: continue playing but receive a painful electric shock, or give up without making it to the end. The shocks were of increasing magnitude and duration. The subjects were divided into two groups, and both were briefed before the game began. One group was trained to deal with the pain from the increasing shocks by distracting themselves. It's as if you were crossing a swamp, they were told, and the best way to cope with it is to imagine a pleasant scene instead. The other group had an equal degree of training; however, this group was trained in acceptance. They were taught that it is possible to continue enduring pain without fighting it, even though it will be getting increasingly intense. They were also given the crossing-a-swamp metaphor. But instead of the suggestion to imagine something pleasant, they were told that the best way to handle the adversity is to notice and accept unpleasant thoughts and not to fight them or the feelings they cause.

The subjects who practiced acceptance proved much better able to go on; that is, they played significantly longer before quitting. Such triumphs are a classic case of rationality and emotion working together, through brain processes the Stoics might have intuited but could not have explained: the executive network structures in our prefrontal cortex exerting influence over the many subcortical structures associated with emotion.[15] When we successfully orchestrate that, we achieve emotion regulation.

REAPPRAISAL: THE POWER OF SPIN

Imagine you're driving to a business meeting and you run into a street blocked by construction. You get lost trying to follow the detour and end up twenty minutes late. You might respond by thinking, "Why can't those idiots provide clear directions!" Such thoughts could make you angry. Alternatively, you might blame yourself, thinking, "Why am I always getting lost? What's wrong with me?" That response might make you frustrated. Or you might respond by thinking about how everyone at the meeting is going to be annoyed with you for being late, and that will make you feel anxious. Each of those negative appraisals of the roadblock and its consequences probably has a bit of truth to it, and, chances are, one of these (or some other) interpretations will be dominant and determine the emotion you feel.

This is how emotions work. Making sense of what just happened is one of the phases your brain goes through as an emotional reaction develops. Psychologists call it "appraisal." Some appraisal goes on in your unconscious mind, but it also occurs on the conscious level, and that's where you can intervene: If there are different ways of looking at something, which lead to different emotions, why not train yourself to think in the way that leads to the emotion you want? In this case, for example, you might guide yourself to think such thoughts as "People won't care if I'm late, because there are many others at the meeting." Or "This won't bother anyone because they know I am usually on time."

Or "Good thing that construction made me late. It gives me a great excuse to miss the first twenty minutes of a boring meeting." Altering the course of how your brain makes sense of things is a way of short-circuiting the cycle that leads to an unwanted emotion. Psychologists call that guided thinking "reappraisal."

There are emotional reactions that empower you and those that disempower you. Empowering emotions help you discover the lessons of every situation and move toward your goals. Disempowering interpretations tie you to negativity and get in the way of your goals. Reappraisal involves recognizing the negative pattern developing in your thoughts and changing it to one that is more desirable, but in a manner that is still based in reality.

Research on reappraisal has shown that we have the power to choose the meanings we assign to circumstances, events, and experiences in our lives. Instead of resenting that server who seems to be ignoring you, you can view your server as a victim of too many tables. Instead of viewing the person who is always boasting about how much money he makes as obnoxious, you can view him as insecure because everyone else in your social group has a more interesting job than he does. Even if the negative appraisals won't completely dissipate, the positive ones add new possibilities to your thinking, moderating the tendency to look at things in a negative way.

One example of the power of reappraisal comes from a recent study by members of the cognitive science team at the U.S. Army Natick Soldier Systems Center (NSSC) in Natick, Massachusetts.[16] The researchers studied twenty-four healthy young people. They had them visit the research lab three times, each time completing a grueling ninety-minute treadmill run. Thirty and sixty minutes into the run, and again when it was over, they were asked about the degree of exertion they felt and any pain or unpleasantness.

The runners were given no instructions regarding how to cope with their first treadmill run. On the next two runs, half the subjects were told to apply cognitive reappraisal to attempt to mitigate their negative feelings—for example, to focus on the cardiac benefit of the exercise or the pride they could take in completing it. The other half, the control group, were told to use a distraction strategy similar to the one used in the acceptance study, such as imagining they were lying on a beach somewhere. The researchers found that, as expected, distraction didn't work, but the group applying reappraisal reported significantly lower levels of exertion and unpleasantness.

Skill at reappraisal doesn't just lead to a more pleasant existence; it can also be the key to workplace success. Because emotions tune your mental calculations, being able to mitigate intense emotion is crucial in many high-pressure professions. Consider a case study led by Mark Fenton-O'Creevy, a professor at the Open University Business School in Milton Keynes, northeast of Oxford.[17]

Soft-spoken, white-haired, and balding, Fenton-O'Creevy has had a varied career: he's worked as a school groundskeeper, a chef, a mathematician in a government research establishment, an outdoor activities instructor, a mathematics teacher, a therapist with emotionally disturbed adolescents, and a management consultant before becoming a business school academic. In 2010 he and some colleagues immersed themselves in the real world of London investment banks to explore the role of emotion and emotion regulation strategies. Due to their collective background, they were able to obtain significant access to a large and high-powered group of finance professionals.

The researchers conducted extensive interviews with 118 professional traders and ten senior managers at four investment banks, three American and one European. Their subjects constituted a representative sample of traders in stocks, bonds, and derivatives. All agreed to report their levels of experience and salary, which according to the compensation plans reflect each subject's degree of trading success. Experience levels ranged from six months to thirty years, and the salaries (including bonus) from about $100,000 to $1 million per year.

Psychologists talk about decision making as proceeding by two parallel processes—the "system one" and "system two" that the Nobel Prize winner Daniel Kahneman popularized in his book *Thinking, Fast and Slow.*[18] System one is fast, based in the unconscious, and capable of processing large amounts of complex material. System two, conscious deliberation, is slow and limited with regard to the amount of information it can take into account at a given time. It is also subject to mental exhaustion.

In the complex and hectic world of securities trading, system one processing is crucial to success, for the fast and sophisticated information flow is beyond the capacity of the conscious mind alone. Just as a baseball player cannot depend on his conscious control to swing his bat and hit a small ball approaching him at ninety miles an hour, so too must traders rely on their unconscious to play a leading role in guiding their decisions.

That's where emotions enter the picture. At the unconscious level, emotions, which draw on past experience, provide a radar that directs your attention and shapes your perception of both threats and opportunities. Through emotion, the steady stream of data and outcomes you've encountered over time will shape your intuition and allow you to rapidly choose the appropriate action.

Consider the role that disgust plays in encoding your experience of foods that could sicken you. If you're about to slurp down an oyster and you notice worms crawling all over it, you don't stop and consciously analyze the details of that situation in light of similar circumstances that you've been in or heard about; you just gag in disgust and throw it down. Similarly, the emotions of traders encapsulate their past trading experience. "People think if you have a PhD you will be very good because you have an understanding of options theory, but this is not always the case," said one manager the researchers interviewed. "You have to also have good gut instincts."

That's the upside of emotion in decision making. The downside shows itself when emotion runs amok. The Fenton-O'Creevy team found that the traders who were least successful, often those with little experience, were the ones who had difficulty keeping their emotions under control.

Trading is a fast-paced and demanding profession, requiring that complex and important decisions be made very quickly.

There is a lot at stake. "Emotionally, it was not easy to cope with," said one trader. "There were times when the desk was down close to $100m." Another conceded, "When you lose money you could sit down and cry. The highs and lows of a trader's life are euphoria or absolute dismay." A third remarked, "There were situations when I was extremely stressed up and then you feel physically ill." Though they obviously struggled with emotion, these and other relatively less successful traders generally denied that emotion played any significant role in their job. They tried to suppress their emotions while at the same time denying that emotions had an effect on their decision making.

The most successful traders had a markedly different attitude. They acknowledged their emotion and showed a great willingness to reflect on their emotion-driven behavior. They recognized that emotion and good decision making are inextricably linked. Accepting that emotions were necessary for high performance, they "tended to reflect critically about the origin of their intuitions and the role of emotion." They embraced the positive and essential role emotions played while understanding that when emotions become too intense, it is useful to know how to tone them down. The issue for the successful traders was not how to avoid emotion but how to regulate and harness it.

Fenton-O'Creevy noted that the most successful emotion regulation approach applied by the traders was reappraisal. If they had a big loss, they might tell themselves that that was to be expected from time to time. Or that just as one big trade is not going to make anyone, one big loss doesn't destroy you; they've all seen fellow traders' fortunes go up and down and how a bad spell was not the end of their world.

The traders' managers recognized the importance of emotion and of effective regulation. Said one, "I have to play the role of director of emotion." But we don't need a boss to do this for us; we can do it ourselves. The first and most crucial step is self-awareness. We all have the capacity to recognize and monitor our feelings. Most people, once they focus on it, realize they are better at it than they'd expected. Then, once we're in touch with our true feelings, we can take steps to manage them by employing the strategies I've been talking about. If we nurture and develop those aspects of emotional intelligence, we can become our own director of emotion, using reappraisal as one of the critical weapons in our arsenal of effective regulation.

EXPRESSION: THE POWER OF WORDS

Karen S. is the COO of a midsized Hollywood production company. It's a demanding and competitive business. Her job requires her to deal with many difficult people, and to be successful, she often has to maintain a good working relationship with clients even after they break commitments or treat her unfairly. At times she grows angry, and that used to get in the way of her work. Then she discovered a remedy: she'll write an email to the offender describing in detail the perceived injustice and openly stating her honest and uncensored feelings about it. But she doesn't send the email. She saves it to her drafts, promising herself to look at it again in a couple days, which she never does. She found that the simple act of expressing her feelings has already solved her problem. She soon gets over the debilitating anger and goes back to work.

Does talking or writing about an emotion help you get over it? Most people are familiar with this approach, but surveys taken by research psychologists show that most people think it doesn't work.[19] On the contrary, they believe that talking amplifies emotion. Among men the willingness to express feelings is especially low. Though, as babies, boys are more socially oriented than girls—they're more likely to look at their mother and display facial expressions of anger or joy—by the time they reach fifteen or sixteen, many males of our species succumb to gender stereotypes and avoid voicing what they feel.[20]

Contrary to public opinion, expressing unwanted negative emotion does help defuse it. Clinical psychologists have found that talking is most effective when the sharing is done with trusted friends or a significant other, especially if those people have experienced similar issues. Finding the right time to talk is also important. Exposing your feelings is important but can be scary, and things can go wrong if the person on the listening end is distracted or doesn't have time to hear you out.

Research psychologists don't have the same hands-on experience as the clinicians, but they've conducted many academic studies of whether and why such talks are beneficial. In the research world talking or writing about your feelings is called "affect labeling."

In recent studies it's been shown that affect labeling has such wide and diverse effects as lowering the distress felt after viewing disturbing photos and videos, calming the anxiety of people who are nervous about public speaking, and reducing the severity of post-traumatic stress disorder. Talking about your feelings increases brain activity in the prefrontal cortex and decreases activity in the amygdala, an effect similar to what is seen when people apply the regulation method of reappraisal.[21] And simply writing about upsetting experiences as Karen S. did has been shown to lower high blood pressure, lessen chronic pain symptoms, and boost immune function.

The benefits of expressing an upsetting emotion can be long lasting. I experienced that myself recently after I was stopped at a red light and a taxi traveling full speed plowed into my vehicle from behind, totaling it, and nearly totaling me. Afterward, I found it uncomfortable to drive. I felt wary, as if, again from nowhere and without warning, another car might careen into me. I was particularly anxious when stopped at a light on a busy street. But as I talked about the accident with friends and acquaintances, I found that sharing my feelings made them fade. The exchanges didn't just calm me at that moment; they had a long-term effect, helping me gradually get over the trauma.

Though there is much anecdotal evidence about the value of talking, and clinicians swear by it, until recently all the scientific studies supporting the benefits of affect labeling have been carried out in psychology labs rather than "in vivo," in people's homes or workplaces. That changed in 2019 when an exciting real-world study by a group of seven scientists appeared in one of the prestigious *Nature* journals.[22]

The scientists studied the emotion displayed in Twitter time lines. Whereas studies in the lab are limited to a few dozen or a few hundred subjects, these researchers were able to analyze the emotional content of twelve-hour streams of tweets from 109,943 Twitter users. The tweets constituted the subjects' real-life thoughts, responses to whatever was happening in their world, captured and preserved in the Twitter servers.

How do you analyze the emotion in more than a million hours of tweets? There is a whole industry devoted to automating such investigations. It's called sentiment analysis, and it's employed in marketing and advertising, linguistics, political science, sociology, and many other fields. The idea is that you feed a string of text into a computer and specialized sentiment analysis software evaluates whether the emotional content is positive or negative, as well as how intense the feelings are.

The authors of the *Nature* article used a program called VADER. The program was developed at the Georgia Institute of Technology and validated on thousands of text passages taken from social media, Rotten Tomatoes movie reviews, *New York Times* opinion pieces, online technical product reviews, and other sources. In an overwhelming proportion of those trial passages, VADER responded with the same ratings as trained human raters.

In their analysis the researchers began by examining more than a billion tweets from more than 600,000 users, looking for any that included an unambiguous statement expressing a feeling for example, "I feel sad" or "I feel very happy." The 109,943 subjects they chose for their study were the ones who had made a tweet of that sort. The researchers then obtained, for each of those tweeters, all tweets made in the six hours prior to the expression of emotion and in the six hours following it. They fed those streams of tweets into the VADER software to create a profile of each user's emotional state during that twelve-hour period.

What they found was remarkable. In the case of negative

emotion, the tweets tended to hold steady at some baseline intensity before starting to build negativity quickly in the half hour or
hour prior to the main emotional expression (for example, "I feel
sad"). Presumably, the buildup and climax were a reaction to some
upsetting information or incident. Then, just after the tweeter
had expressed his or her feelings, there was a rapid decline in the
emotion intensity of the later tweets. The tweet had defused the
bad feeling.

For positive emotion, with apparently no defusing needed, the
curve was much flatter. There was still a buildup before the expression of emotion (for example, "I feel very happy"), but there was
no sharp drop, just a slow decay as the tweeter gradually moved
on to other topics.

What anecdotal and laboratory evidence had suggested had
now been verified by monitoring the emotional pulse of a hundred thousand Twitter users. In *Macbeth*, Shakespeare wrote,
"Give sorrow words. The grief that does not speak whispers the
o'er-fraught heart and bids it break."[23] Like all great playwrights,
Shakespeare was also a great psychologist. He knew that the
tweeters who gave their sorrow words would find relief.

THE JOY OF EMOTION

When I was a child, I got in a lot of trouble. Not just for
things I did, but also for acts I was innocent of. "People blame you
for things because you have a bad reputation," my mother would
tell me. "And once you have a bad reputation, it's hard to change
people's minds." In my studies of the science of emotion, I've
often thought about that. Over many centuries of human thought
and scholarship, emotion had been saddled with a bad reputation,
and it was hard to change that. But in recent years, thanks in great
part to advances in neuroscience, scholars have reshaped the way
we view emotion. Today we now know that the instances of emotion being counterproductive are the exception and not the rule.

I hope that in this tour of the new science of emotion, I've been able to debunk the myth of emotion's counterproductivity and identified how it helps us make the most of our available mental resources. Emotion allows us to have flexible responses depending on our physical states and environmental circumstances, works hand in hand with our systems for wanting and liking to motivate our every action, helps us relate to one another and to cooperate, and pushes us to expand our horizons and reach new heights. Working together with our rational mind, emotion shapes virtually every thought we have. It contributes, moment to moment, to all our judgments and decisions, both large and small, from whether to don a jacket before going outside to how to invest for retirement. Without emotion we'd be lost.

Every species has its ecological niche, each optimized for survival and reproduction in some particular environment or environments. Of all species, humans thrive in the greatest varieties of ecosystems. We live in deserts, in rain forests, in the frozen arctic tundra—even in outer space on the International Space Station. Our resilience builds on our mental flexibility, and that is due in great part to our sophisticated emotions.

Our world, wherever and however we live, presents us with a constant set of challenges. To overcome them, we rely on our senses to detect our surroundings and our thinking to process that information in light of our knowledge and experience. A principal way in which that knowledge and past experience enters our thinking is through emotion. You might not engage in a lot of rational analysis regarding the possibility of starting a fire each time you grill meat in your kitchen, but a tinge of fear of fire always colors your thought and actions around the stove, coaxing you toward safer decisions.

Though emotion is part of the human psychological tool kit, there are differences among individuals. Some are more prone toward fear, others less so, and the same is true for happiness and the other emotions. And though emotion evolved for good reason and is usually beneficial, there are times—especially in our settled, modern world—when it proves counterproductive. The message of this book is that you should appreciate and cherish your emotion and get to know your particular emotional profile, for once you are self-aware, you can manage your feelings so that they always work in your favor.

The Goodbye

As I mentioned earlier, for some years my mother, other than being confined to a wheelchair, had been in good health and living contentedly in an assisted living home. I used to see her once or twice a week for a walk and a chocolate milkshake, but when the coronavirus pandemic came in 2020, her home went on lockdown. That new great calamity that she had feared ever since experiencing the Holocaust—another sudden and tragic disruption of society—had finally materialized.

Soon, many of the staff and residents were diagnosed with COVID-19. Then her home called to say my mother was suspected of having it. It appeared that what Hitler couldn't do, what two decades of smoking, three long-ago bouts of cancer, and a fall down a long flight of stairs in a dive restaurant when she was eighty-five could not do, this microscopic bundle of proteins might accomplish.

A few days later my mother's doctor called me to say that my mother had taken a bad turn and was near death. Because my mother was ninety-eight and had some dementia, it was my decision whether to send her to the hospital. If left in her home, she would die in a day or two, the doctor said. If she went to the hospital immediately, she'd have a chance to survive.

My mother considered hospitals torture—the strange environment, the uncomfortable bed, the IV she hates, the catheter she despises, the parade of strangers going in and out, and the absence of the loving caregivers who watch over her at her home. Last time she was hospitalized, she became agitated and tried to climb out of bed to escape. I had to gather her up and hold her tight to comfort her until she calmed down. This time I would not be allowed to visit her. Could I send her to the hospital for what would probably be a drawn-out and torturous death—in which she would, for all practical purposes, die alone?

Even if my mother did not always have a good life, I felt she deserved a good death. At her home I could see her through a window and tell her I love her. I could let her know that when the end came, even if I wasn't with her in the room, my spirit would be there, hugging her, remembering the times that she helped me after I fell or had a fight at school. I wanted her to feel that I was with her in spirit, holding her hand, kissing her, until her last breath. But if I kept her at the home where I could do that, and where she was happy and comfortable, I would be sentencing her to a certain death. What if the hospital could have saved her?

The doctor said she needed my decision before 6:00 p.m., when she had to leave to do her rounds at the hospital. That meant I had eight minutes. I choked up. My eyes filled with tears. I felt myself tremble. I had trouble thinking logically. I had trouble thinking at all. Do I condemn my mother to death? I can't. Do I condemn my mother to torture? I can't. Having spent so much time researching and writing this book, I knew that emotions are mental states that guide our thoughts, calculations, and decisions, but my emotions didn't seem to be guiding me; they were giving me whiplash.

I asked the doctor if I could think about it and call her back. She hesitated but agreed, warning me that after she left, it would be hard to reach her, so not to call back by 6:00 would mean I was leaving my mother to die in her home.

My son Nicolai once told me I was the most even-keeled, unflappable person he knew. I was proud that I had long ago learned the emotion regulation skills that helped me when I had conflicts with my kids or in my professional life or when investments went bad. But this time, I couldn't get control. I'd shudder thinking of having her sent to the hospital. Then I'd cry, thinking of not sending her.

I felt inadequate. Here I was, in the middle of writing a chapter on how to regulate intense emotion, and in this time of crisis I was collapsing into a puddle of tears. It was now 5:58. I had to give the doctor my decision. I didn't have one, but I didn't want the doctor leaving before I called.

I remembered the study of the securities traders. How the unsuccessful or inexperienced traders tried not to feel while the successful experienced ones accepted their emotions and understood their benefit. Acceptance was what I needed. I needed to let myself feel. I needed to stop fighting my emotions and let them guide me, to let them take the lead. This was too complex and fast a decision for cold hard reason. It was not a decision for my mind; it was one that could be made only by my heart.

I found myself calling the doctor to give her my answer even though I didn't know what my answer would be. As the phone rang and rang, a decision crystallized: I would have my mother stay in her home to die in peace. The doctor finally answered. She asked what I wanted her to do. I told her to send my mother to the hospital.

Like my father watching his fellow underground fighters drive off in that truck, I had decided to do one thing and then done another. The flip-flop surprised me, but I didn't fight it. The doctor told me she thought I'd made the right decision and told the home to call for an ambulance.

My mother did well at the hospital. I was able to speak with her over FaceTime. A busy nurse's aide had to track down the ward's single iPhone and don special protective gear in order for me to do it, but we did that every few days. The nurses reported that my mother didn't suffer as she had in times past, and she responded well to the treatments she was receiving. I was thankful that I hadn't deprived her of her chance to survive. After a week and a half they were ready to send her home. Her doctor marveled at her strength. We all called her the Eveready Bunny.

My mother's home wasn't ready to receive her, however. They were overwhelmed with COVID-19 cases, they said, and had a quota governing how many patients they could take back each day. There was a waiting list. And so my mother stayed in the hospital another day, and then another, and another. At least it was going well; reports were that she still wasn't suffering.

Just as it appeared her home was finally ready to receive her, my mother took a sudden turn for the worse. They changed their minds about sending her home. They were worried about her. She was on heavy oxygen and could no longer talk to me over the phone. That's how it stood when I completed this book. It was on a Friday night, shortly before midnight. I emailed the manuscript to my editor, had a drink, and went to sleep.

A little after 3:00 a.m. I was awakened by a phone call. It was the hospital. My mother had just died.

A couple months have now passed, and I'm finally revising, in these pages, the end to my mother's story. It still hurts to think about it. To imagine her dying without those she loved surrounding her. But I don't regret my decision to send her to the hospital. I'm glad I listened to my heart. I see now that at least it gave her a fighting chance, and I'd never have forgiven myself if I thought I had deprived her of one.

Understanding how your mind and emotions work, then using the knowledge you've gained to manage your emotions more effectively, is not just a science but an art. My friend Deepak Chopra is a master meditator and seems to be able to take any news with equanimity. I suppose he gets that from his meditation. Studies show that meditation produces brain changes that increase executive function and aid you in successfully applying whatever emotion control technique you choose. I have a long way to go on that journey. Writing this book has helped. It has helped me to understand myself, forced me to focus on my emotional life, and taught me many lessons. I hope reading it has had that benefit for you. But there are no miracles. Improving yourself takes constant work and effort, and there will always be situations you wish you could have handled better. The understanding provided by science can help you move on from those disappointments, better armed with the self-knowledge that may prevent such lapses in the future. When they do occur, however, as they will, you can take comfort in knowing that none of us are perfect.

Acknowledgments

This is my eleventh nonfiction book. Some of those who have aided me are the usual suspects and some are new advisers, but one thing all my books share is a large debt to others. Here, I owe the most by far to my great friend, the Caltech neuroscientist Ralph Adolphs. Over the years I worked on *Emotional*, Ralph explained numerous concepts, connected me with other experts, read my drafts, and provided enormous encouragement. His colleague David Anderson was also extremely helpful, as were fellow neuroscientists/psychologists James Russell, James Gross, and Lisa Feldman Barrett. I was also fortunate to be able to pick the brains of two practicing clinical psychologists, Liz Von Schlegel and Kimberly Andersen, and forensic psychiatrist Greg Cohen. Philosopher Nathan King provided insight into the thoughts of the ancient Greeks. My friends and family read numerous drafts and told me where my prose was malfunctioning: Cecilia Milan, Alexei Mlodinow, Nicolai Mlodinow, Olivia Mlodinow, Sanford Perliss, Fred Rose, and my wife, Donna Scott, who is not just loving and supportive but also an amazing editor, whose opinion I treasure and whom I rely upon heavily for advice in all things. I'm also grateful to Andrew Weber at Pantheon Books, and my editor, Edward Kastenmeier, who held me to Pantheon's usual high standard and provided brilliant and constructive advice. I've always felt privileged to enjoy the opportunity to draw on Edward's unsurpassed literary skills and experience, and this book was no exception. Catherine Bradshaw and Susan Ginsburg of Writers House were also always there for me, from the earliest conception of the idea, to the discussions regarding the jacket art. I first met Susan in 2000, and it was the beginning of a beautiful friendship, and a satisfying writing career. Finally, a last goodbye to my dear mother, whose life, and now whose death, provided me with so many lessons that I've drawn upon in my books.

Notes

INTRODUCTION

1. Some neural patterns don't even involve the amygdala. See Justin S. Feinstein et al., "Fear and Panic in Humans with Bilateral Amygdala Damage," *Nature Neuroscience* 16 (2013): 270. For fear and anxiety, see Lisa Feldman Barrett, *How Emotions Are Made* (New York: Houghton Mifflin Harcourt, 2017).
2. Andrew T. Drysdale et al., "Resting-State Connectivity Biomarkers Define Neurophysiological Subtypes of Depression," *Nature Medicine* 23 (2017): 28–38.
3. James Gross and Lisa Feldman Barrett, "The Emerging Field of Affective Neuroscience," *Emotion* 13 (2013): 997–98.
4. James A. Russell, "Emotion, Core Affect, and Psychological Construction," *Cognition and Emotion* 23 (2009): 1259–83.
5. Ralph Adolphs and David J. Anderson, *The Neuroscience of Emotion: A New Synthesis* (Princeton, N.J.: Princeton University Press, 2018), 3.
6. Feldman Barrett, *How Emotions Are Made*, xv.

1
THOUGHT VERSUS FEELING

1. Charlie Burton, "After the Crash: Inside Richard Branson's $600 Million Space Mission," *GQ*, July 2017.
2. Interview with a Scaled Composites employee, Mojave, Calif., Sept. 30, 2017. The interviewee wished to remain anonymous.
3. Melissa Bateson et al., "Agitated Honeybees Exhibit Pessimistic Cognitive Biases," *Current Biology* 21 (2011): 1070–73.
4. Thomas Dixon, " 'Emotion': The History of a Keyword in Crisis," *Emotion Review* 4 (Oct. 2012): 338–44; Tiffany Watt Smith, *The Book of Human Emotions* (New York: Little, Brown, 2016), 6–7.

5. Thomas Dixon, *The History of Emotions Blog*, April 2, 2020, emotionsblog.history.qmul.ac.uk.

6. Amy Maxmen, "Sexual Competition Among Ducks Wreaks Havoc on Penis Size," *Nature* 549 (2017): 443.

7. Kate Wong, "Why Humans Give Birth to Helpless Babies," *Scientific American*, Aug. 28, 2012.

8. Lisa Feldman Barrett, *How Emotions Are Made* (New York: Houghton Mifflin Harcourt, 2017), 167.

9. Ibid., 164–65.

10. See chapter 9 of Rand Swenson, *Review of Clinical and Functional Neuroscience*, Dartmouth Medical School, 2006, www.dartmouth.edu.

11. Peter Farley, "A Theory Abandoned but Still Compelling," *Yale Medicine* (Autumn 2008).

12. Michael R. Gordon, "Ex-Soviet Pilot Still Insists KAL 007 Was Spying," *New York Times*, Dec. 9, 1996.

2

THE PURPOSE OF EMOTION

1. See, for example, Ellen Langer et al., "The Mindlessness of Ostensibly Thoughtful Action: The Role of 'Placebic' Information in Interpersonal Interaction," *Journal of Personality and Social Psychology* 36 (1978): 635–42.

2. "Black Headed Cardinal Feeds Goldfish," YouTube, July 25, 2010, www.youtube.com.

3. Yanfei Liu and K. M. Passino, "Biomimicry of Social Foraging Bacteria for Distributed Optimization: Models, Principles, and Emergent Behaviors," *Journal of Optimization Theory and Applications* 115 (2002): 603–28.

4. Paul B. Rainey, "Evolution of Cooperation and Conflict in Experimental Bacterial Populations," *Nature* 425 (2003): 72; R. Craig MacLean et al., "Evaluating Evolutionary Models of Stress-Induced Mutagenesis in Bacteria," *Nature Reviews Genetics* 14 (2013): 221; Ivan Erill et al., "Aeons of Distress: An Evolutionary Perspective on the Bacterial SOS Response," *FEMS Microbiology Reviews* 31 (2007): 637–56.

5. Antonio Damasio, *The Strange Order of Things: Life, Feeling, and the Making of Cultures* (New York: Pantheon, 2018), 20.

6. Jerry M. Burger et al., "The Pique Technique: Overcoming Mindlessness or Shifting Heuristics?," *Journal of Applied Social Psychology* 37 (2007): 2086–96; Michael D. Santos et al., "Hey Buddy, Can You Spare Seventeen Cents? Mindful Persuasion and the Pique Technique," *Journal of Applied Social Psychology* 24, no. 9 (1994): 755–64.

7. Richard M. Young, "Production Systems in Cognitive Psychology," in *International Encyclopedia of the Social and Behavioral Sciences* (New York: Elsevier, 2001).

8. F. B. M. de Waal, *Chimpanzee Politics: Power and Sex Among Apes* (Baltimore: Johns Hopkins University Press, 1982).

9. Interview with Anderson, June 13, 2018.

10. Kaspar D. Mossman, "Profile of David J. Anderson," *PNAS* 106 (2009): 17623–25.

11. Yael Grosjean et al., "A Glial Amino-Acid Transporter Controls Synapse Strength and Homosexual Courtship in Drosophila," *Nature Neuroscience* 11, no. 1 (2008): 54–61.

12. G. Shohat-Ophir et al., "Sexual Deprivation Increases Ethanol Intake in Drosophila," *Science* 335 (2012): 1351–55.

13. Paul R. Kleinginna and Anne M. Kleinginna, "A Categorized List of Emotion Definitions, with Suggestions for a Consensual Definition," *Motivation and Emotion* 5 (1981): 345–79. See also Carroll E. Izard, "The Many Meanings/Aspects of Emotion: Definitions, Functions, Activation, and Regulation," *Emotion Review* 2 (2010): 363–70.

14. The correct technical term is "reinforcing."

15. Stephanie A. Shields and Beth A. Koster, "Emotional Stereotyping of Parents in Child Rearing Manuals, 1915–1980," *Social Psychology Quarterly* 52, no. 1 (1989): 44–55.

3

THE MIND-BODY CONNECTION

1. W. B. Cannon, *The Wisdom of the Body* (New York: W. W. Norton, 1932).

2. See, for example, James A. Russell, "Core Affect and the Psycho-

logical Construction of Emotion," *Psychological Review* 110 (2003): 145–72; Michelle Yik, James A. Russell, and James H. Steiger, "A 12-Point Circumplex Structure of Core Affect," *Emotion* 11 (2011): 705. See also Antonio Damasio, *The Strange Order of Things: Life, Feeling, and the Making of Cultures* (New York: Pantheon, 2018). There, Damasio describes what is essentially the effect of core affect, which he calls homeostatic feeling.

3. Christine D. Wilson-Mendenhall et al., "Neural Evidence That Human Emotions Share Core Affective Properties," *Psychological Science* 24 (2013): 947–56.

4. Ibid.

5. Michael L. Platt and Scott A. Huettel, "Risky Business: The Neuro-economics of Decision Making Under Uncertainty," *Nature Neuro-science* 11 (2008): 398–403; Thomas Caraco, "Energy Budgets, Risk, and Foraging Preferences in Dark-Eyed Juncos (*Junco hyemalis*)," *Behavioral Ecology and Sociobiology* 8 (1981): 213–17.

6. John Donne, *Devotions upon Emergent Occasions* (Cambridge, U.K.: Cambridge University Press, 2015), 98.

7. Damasio, *Strange Order of Things*, chap. 4.

8. Shadi S. Yarandi et al., "Modulatory Effects of Gut Microbiota on the Central Nervous System: How Gut Could Play a Role in Neuropsychiatric Health and Diseases," *Journal of Neurogastroenterology and Motility* 22 (2016): 201.

9. Tal Shomrat and Michael Levin, "An Automated Training Paradigm Reveals Long-Term Memory in Planarians and Its Persistence Through Head Regeneration," *Journal of Experimental Biology* 216 (2013): 3799–810.

10. Stephen M. Collins et al., "The Adoptive Transfer of Behavioral Phenotype via the Intestinal Microbiota: Experimental Evidence and Clinical Implications," *Current Opinion in Microbiology* 16, no. 3 (2013): 240–45.

11. Peter Andrey Smith, "Brain, Meet Gut," *Nature* 526, no. 7573 (2015): 312.

12. See, for example, Tyler Halverson and Kannayiram Alagiakrishnan, "Gut Microbes in Neurocognitive and Mental Health Disorders," *Annals of Medicine* 52 (2020): 423–43.

13. Gale G. Whiteneck et al., *Aging with Spinal Cord Injury* (New York: Demos Medical Publishing, 1993), vii.

14. George W. Hohmann, "Some Effects of Spinal Cord Lesions on Experienced Emotional Feelings," *Psychophysiology* 3 (1966): 143–56.

15. See, for example, Francesca Pistoia et al., "Contribution of Interoceptive Information to Emotional Processing: Evidence from Individuals with Spinal Cord Injury," *Journal of Neurotrauma* 32 (2015): 1981–86.

16. Nayan Lamba et al., "The History of Head Transplantation: A Review," *Acta Neurochirurgica* 158 (2016): 2239–47.

17. Sergio Canavero, "HEAVEN: The Head Anastomosis Venture Project Outline for the First Human Head Transplantation with Spinal Linkage," *Surgical Neurology International* 4 (2013): S335–S342.

18. Paul Root Wolpe, "A Human Head Transplant Would Be Reckless and Ghastly. It's Time to Talk About It," *Vox*, June 12, 2018, www.vox.com.

19. Rainer Reisenzein et al., "The Cognitive-Evolutionary Model of Surprise: A Review of the Evidence," *Topics in Cognitive Science* 11 (2019): 50–74.

20. Shai Danziger et al., "Extraneous Factors in Judicial Decisions," *Proceedings of the National Academy of Sciences* 108 (2011): 6889–92.

21. Jeffrey A. Linder et al., "Time of Day and the Decision to Prescribe Antibiotics," *JAMA Internal Medicine* 174 (2014): 2029–31.

22. Shai Danziger et al., "Extraneous Factors in Judicial Decisions," *Proceedings of the National Academy of Sciences* 108 (2011): 6889–92.

23. Jing Chen et al., "Oh What a Beautiful Morning! Diurnal Influences on Executives and Analysts: Evidence from Conference Calls," *Management Science* (Jan. 2018).

24. Brad J. Bushman, "Low Glucose Relates to Greater Aggression in Married Couples," *PNAS* 111 (2014): 6254–57.

25. Christina Sagioglou and Tobias Greitemeyer, "Bitter Taste Causes Hostility," *Personality and Social Psychology Bulletin* 40 (2014): 1589–97.

4

HOW EMOTIONS GUIDE THOUGHT

1. Most of the Dirac story is from Graham Farmelo, *The Strangest Man: The Hidden Life of Paul Dirac, Mystic of the Atom* (New York: Perseus, 2009), 252–63.

2. Ibid., 293.

3. Ibid., 438.

4. Barry Leibowitz, "Wis. Man Got Shot—Intentionally—in 'Phenomenally Stupid' Attempt to Win Back Ex-girlfriend," CBS News, July 28, 2011, www.cbsnews.com; Paul Thompson, " 'Phenomenally Stupid' Man Has His Friends Shoot Him Three Times to Win Ex-girlfriend's Pity," *Daily Mail*, July 28, 2011.

5. Interview with Perliss, Perliss Law Center, Dec. 9, 2020.

6. See John Tooby and Leda Cosmides, "The Evolutionary Psychology of the Emotions and Their Relationship to Internal Regulatory Variables," in *Handbook of Emotions*, 3rd ed., eds. Michael Lewis, Jeannette M. Haviland-Jones, and Lisa Feldman Barrett (New York: Guilford, 2008), 114–37.

7. Eric J. Johnson and Amos Tversky, "Affect, Generalization, and the Perception of Risk," *Journal of Personality and Social Psychology* 45 (1983): 20.

8. Aaron Sell et al., "Formidability and the Logic of Human Anger," *Proceedings of the National Academy of Sciences* 106 (2009): 15073–78.

9. Edward E. Smith et al., *Atkinson and Hilgard's Introduction to Psychology* (Belmont, Calif.: Wadsworth, 2003), 147; Elizabeth Loftus, *Witness for the Defense: The Accused, the Eyewitness, and the Expert Who Puts Memory on Trial* (New York: St. Martin's Press, 2015).

10. Michel Tuan Pham, "Emotion and Rationality: A Critical Review and Interpretation of Empirical Evidence," *Review of General Psychology* 11 (2007): 155.

11. Carmelo M. Vicario et al., "Core, Social, and Moral Disgust Are Bounded: A Review on Behavioral and Neural Bases of Repugnance in Clinical Disorders," *Neuroscience and Biobehavioral Reviews* 80 (2017): 185–200; Borg Schaich et al., "Infection, Incest, and Iniquity: Investigating the Neural Correlates of Disgust and Morality," *Journal of Cognitive Neuroscience* 20 (2008): 1529–46.

12. Simone Schnall et al., "Disgust as Embodied Moral Judgment," *Personality and Social Psychology Bulletin* 34 (2008): 1096–109.

13. Kendall J. Eskine et al., "A Bad Taste in the Mouth: Gustatory Disgust Influences Moral Judgment," *Psychological Science* 22 (2011): 295–99.

14. Kendall J. Eskine et al., "The Bitter Truth About Morality: Virtue, Not Vice, Makes a Bland Beverage Taste Nice," *PLoS One* 7 (2012): e41159.

15. Mark Schaller and Justin H. Park, "The Behavioral Immune System (and Why It Matters)," *Current Directions in Psychological Science* 20 (2011): 99–103.

16. Dalvin Brown, "'Fact Is I Had No Reason to Do It': Thousand Oaks Gunman Posted to Instagram During Massacre," *USA Today*, Nov. 10, 2018.

17. Pham, "Emotion and Rationality."

18. See, for example, Ralph Adolphs, "Emotion," *Current Biology* 13 (2010).

19. Alison Jing Xu et al., "Hunger Promotes Acquisition of Nonfood Objects," *Proceedings of the National Academy of Sciences* (2015): 201417712.

20. Seunghee Han et al., "Disgust Promotes Disposal: Souring the Status Quo" (Faculty Research Working Paper Series, RWP10-021, John F. Kennedy School of Government, Harvard University, 2010); Jennifer S. Lerner et al., "Heart Strings and Purse Strings: Carryover Effects of Emotions on Economic Decisions," *Psychological Science* 15 (2004): 337–41.

21. Laith Al-Shawaf et al., "Human Emotions: An Evolutionary Psychological Perspective," *Emotion Review* 8 (2016): 173–86.

22. Dan Ariely and George Loewenstein, "The Heat of the Moment: The Effect of Sexual Arousal on Sexual Decision Making," *Journal of Behavioral Decision Making* 19 (2006): 87–98.

23. See, for instance, Martie G. Haselton and David M. Buss, "The Affective Shift Hypothesis: The Functions of Emotional Changes Following Sexual Intercourse," *Personal Relationships* 8 (2001): 357–69.

24. See, for example, B. Kyu Kim and Gal Zauberman, "Can Victoria's Secret Change the Future? A Subjective Time Perception Account

of Sexual-Cue Effects on Impatience," *Journal of Experimental Psychology: General* 142 (2013): 328.

25. Donald Symons, *The Evolution of Human Sexuality* (New York: Oxford University Press, 1979), 212–13.

26. Shayna Skakoon-Sparling et al., "The Impact of Sexual Arousal on Sexual Risk-Taking and Decision-Making in Men and Women," *Archives of Sexual Behavior* 45 (2016): 33–42.

27. Charmaine Borg and Peter J. de Jong, "Feelings of Disgust and Disgust-Induced Avoidance Weaken Following Induced Sexual Arousal in Women," *PLoS One* 7 (Sept. 2012): 1–7.

28. Hassan H. López et al., "Attractive Men Induce Testosterone and Cortisol Release in Women," *Hormones and Behavior* 56 (2009): 84–92.

29. Sir Ernest Shackleton, *The Heart of the Antarctic* (London: Wordsworth Editions, 2007), 574.

30. Michelle N. Shiota et al., "Beyond Happiness: Building a Science of Discrete Positive Emotions," *American Psychologist* 72 (2017): 617–43.

31. Barbara L. Fredrickson and Christine Branigan, "Positive Emotions Broaden the Scope of Attention and Thought-Action Repertoires," *Cognition and Emotion* 19 (2005): 313–32.

32. Barbara L. Fredrickson, "The Role of Positive Emotions in Positive Psychology: The Broaden-and-Build Theory of Positive Emotions," *American Psychologist* 56 (2001): 218; Barbara L. Fredrickson, "What Good Are Positive Emotions?," *Review of General Psychology* 2 (1998): 300.

33. Paul Piff and Dachar Keltner, "Why Do We Experience Awe?," *New York Times*, May 22, 2015.

34. Samantha Dockray and Andrew Steptoe, "Positive Affect and Psychobiological Processes," *Neuroscience and Biobehavioral Reviews* 35 (2010): 69–75.

35. Andrew Steptoe et al., "Positive Affect and Health-Related Neuroendocrine, Cardiovascular, and Inflammatory Processes," *Proceedings of the National Academy of Sciences* 102 (2005): 6508–12.

36. Sheldon Cohen et al., "Emotional Style and Susceptibility to the Common Cold," *Psychosomatic Medicine* 65 (2003): 652–57.

37. B. Grinde, "Happiness in the Perspective of Evolutionary Psychology," *Journal of Happiness Studies* 3 (2002): 331–54.
38. Chris Tkach and Sonja Lyubomirsky, "How Do People Pursue Happiness? Relating Personality, Happiness-Increasing Strategies, and Well-Being," *Journal of Happiness Studies* 7 (2006): 183–225.
39. Melissa M. Karnaze and Linda J. Levine, "Sadness, the Architect of Cognitive Change," in *The Function of Emotions*, ed. Heather C. Lench (New York: Springer, 2018).
40. Kevin Au et al., "Mood in Foreign Exchange Trading: Cognitive Processes and Performance," *Organizational Behavior and Human Decision Processes* 91 (2003): 322–38.

5

WHERE FEELINGS COME FROM

1. Anton J. M. De Craen et al., "Placebos and Placebo Effects in Medicine: Historical Overview," *Journal of the Royal Society of Medicine* 92 (1999): 511–15.
2. Leonard A. Cobb et al., "An Evaluation of Internal-Mammary-Artery Ligation by a Double-Blind Technic," *New England Journal of Medicine* 260 (1959): 1115–18; E. Dimond et al., "Comparison of Internal Mammary Artery Ligation and Sham Operation for Angina Pectoris," *American Journal of Cardiology* 5 (1960): 483–86.
3. Rasha Al-Lamee et al., "Percutaneous Coronary Intervention in Stable Angina (ORBITA): A Double-Blind, Randomised Controlled Trial," *Lancet* 39 (2018): 31–40.
4. Gina Kolata, "'Unbelievable': Heart Stents Fail to Ease Chest Pain," *New York Times*, Nov. 2, 2017.
5. Michael Boiger and Batja Mesquita, "A Socio-dynamic Perspective on the Construction of Emotion," in *The Psychological Construction of Emotions*, ed. Lisa Feldman Barrett and James A. Russell (New York: Guilford Press, 2015), 377–98.
6. Rainer Reisenstein, "The Schachter Theory of Emotion: Two Decades Later," *Psychological Bulletin* 94 (1983): 239–64; Randall L. Rose and Mandy Neidermeyer, "From Rudeness to Road Rage: The Antecedents and Consequences of Consumer Aggression," in

Advances in Consumer Research, ed. Eric J. Arnould and Linda M. Scott (Provo, Utah: Association for Consumer Research, 1999), 12–17.

7. Richard M. Warren, "Perceptual Restoration of Missing Speech Sounds," *Science*, Jan. 23, 1970, 392–93; Richard M. Warren and Roselyn P. Warren, "Auditory Illusions and Confusions," *Scientific American* 223 (1970): 30–36.

8. Robin Goldstein et al., "Do More Expensive Wines Taste Better? Evidence from a Large Sample of Blind Tastings," *Journal of Wine Economics* 3, no. 1 (Spring 2008): 1–9.

9. William James, "The Physical Basis of Emotion," *Psychological Review* 1 (1894): 516–29.

10. J. S. Feinstein et al., "Fear and Panic in Humans with Bilateral Amygdala Damage," *Nature Neuroscience* 16 (2013): 270–72.

11. Lisa Feldman Barrett, "Variety Is the Spice of Life: A Psychological Construction Approach to Understanding Variability in Emotion," *Cognition and Emotion* 23 (2009): 1284–306.

12. Ibid.

13. Boiger and Mesquita, "Socio-dynamic Perspective on the Construction of Emotion."

14. R. I. Levy, *Tahitians: Mind and Experience in the Society Islands* (Chicago: University of Chicago Press, 1975).

15. James A. Russell, "Culture and the Categorization of Emotions," *Psychological Bulletin* 110 (1991): 426; James A. Russell, "Natural Language Concepts of Emotion," *Perspectives in Personality* 3 (1991): 119–37.

16. Ralph Adolphs et al., "What Is an Emotion?," *Current Biology* 29 (2019): R1060–R1064.

17. David Strege, "Elephant's Road Rage Results in Fatality," *USA Today*, Nov. 30, 2018.

18. Peter Salovey and John D. Mayer, "Emotional Intelligence," *Imagination, Cognition, and Personality* 9 (1990): 185–211.

19. Adam D. Galinsky et al., "Why It Pays to Get Inside the Head of Your Opponent: The Differential Effect of Perspective Taking and Empathy in Strategic Interactions," *Psychological Science* 19 (2008): 378–84.

20. Diana I. Tamir and Jason P. Mitchell, "Disclosing Information

About the Self Is Intrinsically Rewarding," *Proceedings of the National Academy of Sciences* 109 (2012): 8038–43.

6

MOTIVATION: WANTING VERSUS LIKING

1. Sophie Roberts, "You Can't Eat It," *Sun*, May 16, 2017, www.thesun .co.uk.
2. Ella P. Lacey, "Broadening the Perspective of Pica: Literature Review," *Public Health Reports* 105, no. 1 (1990): 29.
3. Tom Lorenzo, "Michel Lotito: The Man Who Ate Everything," CBS Local, Oct. 1, 2012, tailgatefan.cbslocal.com.
4. Junko Hara et al., "Genetic Ablation of Orexin Neurons in Mice Results in Narcolepsy, Hypophagia, and Obesity," *Neuron* 30 (2001): 345–54.
5. Robert G. Heath, "Pleasure and Brain Activity in Man," *Journal of Nervous and Mental Disease* 154 (1972): 3–17.
6. For Heath's story, see Robert Colville, "'The 'Gay Cure' Experiments That Were Written out of Scientific History," *Mosaic*, July 4, 2016, mosaicscience.com; Judith Hooper and Dick Teresi, *The Three-Pound Universe* (New York: Tarcher, 1991), 152–61; Christen O'Neal et al., "Dr. Robert G. Heath: A Controversial Figure in the History of Deep Brain Stimulation," *Neurosurgery Focus* 43 (2017): 1–8; John Gardner, "A History of Deep Brain Stimulation: Technological Innovation and the Role of Clinical Assessment Tools," *Social Studies of Science* 43 (2013): 707–28.
7. Dominik Gross and Gereon Schäfer, "Egas Moniz (1874–1955) and the 'Invention' of Modern Psychosurgery: A Historical and Ethical Reanalysis Under Special Consideration of Portuguese Original Sources," *Neurosurgical Focus* 30, no. 2 (2011): E8.
8. Elizabeth Johnston and Leah Olsson, *The Feeling Brain: The Biology and Psychology of Emotions* (New York: W. W. Norton, 2015), 125; Bryan Kolb and Ian Q. Whishaw, *An Introduction to Brain and Behavior*, 2nd ed. (New York: Worth Publishers, 2004), 392–94; Patrick Anselme and Mike J. F. Robinson, "'Wanting,' 'Liking,' and Their Relation to Consciousness," *Journal of Experimental Psychology: Animal Learning and Cognition* 42 (2016): 123–40.

9. Johnston and Olsson, *Feeling Brain*, 125.
10. Daniel H. Geschwind and Jonathan Flint, "Genetics and Genomics of Psychiatric Disease," *Science* 349 (2015): 1489–94; T. D. Cannon, "How Schizophrenia Develops: Cognitive and Brain Mechanisms Underlying Onset of Psychosis," *Trends in Cognitive Science* 19 (2015): 744–56.
11. Peter Milner, "Peter M. Milner," Society for Neuroscience, www.sfn.org.
12. Lauren A. O'Connell and Hans A. Hofmann, "The Vertebrate Mesolimbic Reward System and Social Behavior Network: A Comparative Synthesis," *Journal of Comparative Neurology* 519 (2011): 3599–639.
13. Anselme and Robinson, "'Wanting,' 'Liking,' and Their Relation to Consciousness," 123–40.
14. Amy Fleming, "The Science of Craving," *Economist*, May 7, 2015; Anselme and Robinson, "'Wanting,' 'Liking,' and Their Relation to Consciousness."
15. Kent C. Berridge, "Measuring Hedonic Impact in Animals and Infants: Microstructure of Affective Taste Reactivity Patterns," *Neuroscience and Biobehavioral Reviews* 24 (2000): 173–98.
16. For a summary of Berridge's early work and ideas, see Terry E. Robinson and Kent C. Berridge, "The Neural Basis of Drug Craving: An Incentive-Sensitization Theory of Addiction," *Brain Research Reviews* 18 (1993): 247–91.
17. Kent C. Berridge and Elliot S. Valenstein, "What Psychological Process Mediates Feeding Evoked by Electrical Stimulation of the Lateral Hypothalamus?," *Behavioral Neuroscience* 105 (1991).
18. Anselme and Robinson, "'Wanting,' 'Liking,' and Their Relation to Consciousness," 123–140; see also Berridge website, and Johnston and Olsson, *Feeling Brain*, 123–43.
19. For a review, see Kent C. Berridge and Morten L. Kringelbach, "Neuroscience of Affect: Brain Mechanisms of Pleasure and Displeasure," *Current Opinion in Neurobiology* 23 (2013): 294–303; Anselme and Robinson, "'Wanting,' 'Liking,' and Their Relation to Consciousness," 123–40.
20. Ab Litt, Uzma Khan, and Baba Shiv, "Lusting While Loathing: Par-

allel Counterdriving of Wanting and Liking," *Psychological Science* 21, no. 1 (2010): 118–25, dx.doi.org/10.1177/0956797609355633.

21. M. J. F. Robinson et al., "Roles of 'Wanting' and 'Liking' in Motivating Behavior: Gambling, Food, and Drug Addictions," in *Behavioral Neuroscience of Motivation*, eds. Eleanor H. Simpson and Peter D. Balsam (New York: Springer, 2016), 105–36.

22. Xianchi Dai, Ping Dong, and Jayson S. Jia, "When Does Playing Hard to Get Increase Romantic Attraction?," *Journal of Experimental Psychology: General* 143 (2014): 521.

23. *The History of Xenophon*, trans. Henry Graham Dakyns (New York: Tandy-Thomas, 1909), 4:64–71.

24. Fleming, "Science of Craving."

25. Anselme and Robinson, " 'Wanting,' 'Liking,' and Their Relation to Consciousness," 123–40.

26. Wilhelm Hofmann et al., "Desire and Desire Regulation," in *The Psychology of Desire*, ed. Wilhelm Hofmann and Loran F. Nordgren (New York: Guilford Press, 2015).

27. Anselme and Robinson, " 'Wanting,' 'Liking,' and Their Relation to Consciousness," 123–40; Todd Love et al., "Neuroscience of Internet Pornography Addiction: A Review and Update," *Behavioral Sciences* 5, no. 3 (2015): 388–433. The nucleus accumbens receives the dopamine signal from the ventral tegmental area. All drugs of abuse affect that "mesolimbic dopamine (DA) pathway," from the ventral tegmental area into the nucleus accumbens.

28. Morton Kringelbach and Kent Berridge, "Motivation and Pleasure in the Brain," in Hofmann and Nordgren, *Psychology of Desire*.

29. Wendy Foulds Mathes et al., "The Biology of Binge Eating," *Appetite* 52 (2009): 545–53.

30. "Sara Lee Corp.," *Advertising Age*, Sept. 2003, adage.com.

31. Paul M. Johnson and Paul J. Kenny, "Addiction-Like Reward Dysfunction and Compulsive Eating in Obese Rats: Role for Dopamine D2 Receptors," *Nature Neuroscience* 13 (2010): 635.

32. For the record, Sara Lee's Classic New York Style Cheesecake contains cream cheese, sugar, eggs, enriched flour, high fructose corn syrup, partially hydrogenated vegetable oil (soybean and/or cottonseed oils), dextrose, maltodextrin, whole wheat flour, water,

cultured skim milk, cream, corn starch, skim milk, salt, leavening (sodium acid pyrophosphate, baking soda, monocalcium phosphate, calcium sulfate), modified corn and tapioca starch, gums (xanthan, carob bean, guar), vanillin, molasses, cinnamon, carrageenan, potassium chloride, soy flour.

33. Michael Moss, "The Extraordinary Science of Addictive Junk Food," *New York Times*, Feb. 20, 2013.

34. Ashley N. Gearhardt et al., "The Addiction Potential of Hyperpalatable Foods," *Current Drug Abuse Reviews* 4 (2011): 140–45.

35. Robinson et al., "Roles of 'Wanting' and 'Liking' in Motivating Behavior."

36. Bernard Le Foll et al., "Genetics of Dopamine Receptors and Drug Addiction: A Comprehensive Review," *Behavioural Pharmacology* 20 (2009): 1–17.

37. Nikolaas Tinbergen, *The Study of Instinct* (New York: Oxford University Press, 1951); Deirdre Barrett, *Supernormal Stimuli: How Primal Urges Overran Their Evolutionary Purpose* (New York: W. W. Norton, 2010).

38. Gearhardt et al., "Addiction Potential of Hyperpalatable Foods."

39. Moss, "Extraordinary Science of Addictive Junk Food."

40. K. M. Flegal et al., "Estimating Deaths Attributable to Obesity in the United States," *American Journal of Public Health* 94 (2004): 1486–89.

7

DETERMINATION

1. The account is from John Johnson and Bill Long, *Tyson-Douglas: The Inside Story of the Upset of the Century* (Lincoln, Neb.: Potomac Books, 2008), and Joe Layden, *The Last Great Fight: The Extraordinary Tale of Two Men and How One Fight Changed Their Lives Forever* (New York: Macmillan, 2008); Martin Domin, "Buster Douglas Reveals His Mum Was the Motivation for Mike Tyson Upset as Former World Champion Recalls Fight 25 Years On," *Mail Online*, Feb. 11, 2015, www.dailymail.co.uk.

2. Muhammad Ali, *The Greatest: My Own Story*, with Richard Durham (New York: Random House, 1975).

3. Martin Fritz Huber, "A Brief History of the Sub-4-Minute Mile," *Outside*, June 9, 2017, www.outsideonline.com.
4. William Shakespeare, *The Tragedy of Hamlet, Prince of Denmark*, act 3, scene 1.
5. David D. Daly and J. Grafton Love, "Akinetic Mutism," *Neurology* 8 (1958).
6. William W. Seeley et al., "Dissociable Intrinsic Connectivity Networks for Salience Processing and Executive Control," *Journal of Neuroscience* 27 (2007): 2349–56.
7. Emily Singer, "Inside a Brain Circuit, the Will to Press On," *Quanta Magazine*, Dec. 5, 2013, www.quantamagazine.org.
8. Josef Parvizi et al., "The Will to Persevere Induced by Electrical Stimulation of the Human Cingulate Gyrus," *Neuron* 80 (2013): 1259–367.
9. Singer, "Inside a Brain Circuit, the Will to Press On."
10. Erno J. Hermans et al., "Stress-Related Noradrenergic Activity Prompts Large-Scale Neural Network Reconfiguration," *Science* 334 (2011): 1151–53; Andrea N. Goldstein and Matthew P. Walker, "The Role of Sleep in Emotional Brain Function," *Annual Review of Clinical Psychology* 10 (2014): 679–708.
11. Tingting Zhou et al., "History of Winning Remodels Thalamo-PFC Circuit to Reinforce Social Dominance," *Science* 357 (2017): 162–68.
12. See, for example, M. C. Pensel et al., "Executive Control Processes Are Associated with Individual Fitness Outcomes Following Regular Exercise Training: Blood Lactate Profile Curves and Neuroimaging Findings," *Science Reports* 8 (2018): 4893; S. F. Sleiman et al., "Exercise Promotes the Expression of Brain Derived Neurotrophic Factor (BDNF) Through the Action of the Ketone Body β-hydroxybutyrate," *eLife* 5 (2016): e15092.
13. Y. Y. Tang et al., "Brief Meditation Training Induces Smoking Reduction," *Proceedings of the National Academy of Sciences, USA* 110 (2013): 13971–75.
14. Robert S. Marin, Ruth C. Biedrzycki, and Sekip Firinciogullari, "Reliability and Validity of the Apathy Evaluation Scale," *Psychiatry Research* 38 (1991): 143–62; Robert S. Marin and Patricia A. Wilkosz, "Disorders of Diminished Motivation," *Journal of Head*

Trauma Rehabilitation 20 (2005): 377–88; Brendan J. Guercio, "The Apathy Evaluation Scale: A Comparison of Subject, Informant, and Clinician Report in Cognitively Normal Elderly and Mild Cognitive Impairment," *Journal of Alzheimer's Disease* 47 (2015): 421–32; Richard Levy and Bruno Dubois, "Apathy and the Functional Anatomy of the Prefrontal Cortex–Basal Ganglia Circuits," *Cerebral Cortex* 16 (2006): 916–28.

15. Goldstein and Walker, "Role of Sleep in Emotional Brain Function."

16. Ibid.

17. Matthew Walker, *Why We Sleep: Unlocking the Power of Sleep and Dreams* (New York: Scribner, 2017), 204.

8

YOUR EMOTIONAL PROFILE

1. See, for example, Richard J. Davidson, "Well-Being and Affective Style: Neural Substrates and Biobehavioural Correlates," *Philosophical Transactions of the Royal Society of London, Series B: Biological Sciences* 359 (2004): 1395–411.

2. Mary K. Rothbart, "Temperament, Development, and Personality," *Current Directions in Psychological Science* 16 (2007): 207–12.

3. Richard J. Davidson and Sharon Begley, *The Emotional Life of Your Brain* (New York: Plume, 2012), 97–102.

4. Greg Miller, "The Seductive Allure of Behavioral Epigenetics," *Science* 329 (2010): 24–29.

5. June Price Tangney and Ronda L. Dearing, *Shame and Guilt* (New York: Guilford Press, 2002), 207–14.

6. See, for example, the control group results in Giorgio Coricelli, Elena Rusconi, and Marie Claire Villeval, "Tax Evasion and Emotions: An Empirical Test of Re-integrative Shaming Theory," *Journal of Economic Psychology* 40 (2014): 49–61; Jessica R. Peters and Paul J. Geiger, "Borderline Personality Disorder and Self-Conscious Affect: Too Much Shame but Not Enough Guilt?," *Personality Disorders: Theory, Research, and Treatment* 7, no. 3 (2016): 303; Kristian L. Alton, "Exploring the Guilt-Proneness of Nontraditional Students" (master's thesis, Southern Illinois University

at Carbondale, 2012); Nicolas Rüsch et al., "Measuring Shame and Guilt by Self-Report Questionnaires: A Validation Study," *Psychiatry Research* 150, no. 3 (2007): 313–25.

7. Tangney and Dearing, *Shame and Guilt*.

8. See, for example, Souheil Hallit et al., "Validation of the Hamilton Anxiety Rating Scale and State Trait Anxiety Inventory A and B in Arabic Among the Lebanese Population," *Clinical Epidemiology and Global Health* 7 (2019): 464–70; Ana Carolina Monnerat Fioravanti-Bastos, Elie Cheniaux, and J. Landeira-Fernandez, "Development and Validation of a Short-Form Version of the Brazilian State-Trait Anxiety Inventory," *Psicologia: Reflexão e Crítica* 24 (2011): 485–94.

9. Konstantinos N. Fountoulakis et al., "Reliability and Psychometric Properties of the Greek Translation of the State-Trait Anxiety Inventory Form Y: Preliminary Data," *Annals of General Psychiatry* 5, no. 2 (2006): 6.

10. See, for example, ibid.; Tracy A. Dennis, "Interactions Between Emotion Regulation Strategies and Affective Style: Implications for Trait Anxiety Versus Depressed Mood," *Motivation and Emotion* 31 (2007): 203.

11. Arnold H. Buss and Mark Perry, "The Aggression Questionnaire," *Journal of Personality and Social Psychology* 63 (1992): 452–59.

12. Judith Orloff, *Emotional Freedom* (New York: Three Rivers Press, 2009), 346.

13. Peter Hills and Michael Argyle, "The Oxford Happiness Questionnaire: Compact Scale for the Measurement of Psychological Well-Being," *Personality and Individual Differences* 33 (2002): 1073–82.

14. Mean scores on the Oxford Happiness Questionnaire were surprisingly similar in studies across different professions and the globe. See, for example, Ellen Chung, Vloreen Nity Mathew, and Geetha Subramaniam, "In the Pursuit of Happiness: The Role of Personality," *International Journal of Academic Research in Business and Social Sciences* 9 (2019): 10–19; Nicole Hadjiloucas and Julie M. Fagan, "Measuring Happiness and Its Effect on Health in Individuals That Share Their Time and Talent While Participating in 'Time Banking'" (2014); Madeline Romaniuk, Justine Evans, and Chloe Kidd, "Evaluation of an Equine-Assisted Therapy Program for Veterans Who Identify as 'Wounded, Injured, or Ill' and Their Partners,"

PLoS One 13 (2018); Leslie J. Francis and Giuseppe Crea, "Happiness Matters: Exploring the Linkages Between Personality, Personal Happiness, and Work-Related Psychological Health Among Priests and Sisters in Italy," *Pastoral Psychology* 67 (2018): 17–32; Mandy Robbins, Leslie J. Francis, and Bethan Edwards, "Prayer, Personality, and Happiness: A Study Among Undergraduate Students in Wales," *Mental Health, Religion, and Culture* 11 (2008): 93–99.

15. Ed Diener et al., "Happiness of the Very Wealthy," *Social Indicators Research* 16 (1985): 263–74.

16. Kennon M. Sheldon and Sonja Lyubomirsky, "Revisiting the Sustainable Happiness Model and Pie Chart: Can Happiness Be Successfully Pursued?," *Journal of Positive Psychology* (2019): 1–10.

17. Sonja Lyubomirsky, *The How of Happiness: A Scientific Approach to Getting the Life You Want* (New York: Penguin Press, 2008).

18. R. Chris Fraley, "Information on the Experiences in Close Relationships-Revised (ECR-R) Adult Attachment Questionnaire," labs.psychology.illinois.edu.

19. Semir Zeki, "The Neurobiology of Love," *FEBS Letters* 581 (2007): 2575–79.

20. T. Joel Wade, Gretchen Auer, and Tanya M. Roth, "What Is Love: Further Investigation of Love Acts," *Journal of Social, Evolutionary, and Cultural Psychology* 3 (2009): 290.

21. Piotr Sorokowski et al., "Love Influences Reproductive Success in Humans," *Frontiers in Psychology* 8 (2017): 1922.

22. Jeremy Axelrod, "Philip Larkin: 'An Arundel Tomb,'" www.poetryfoundation.org.

9

MANAGING EMOTIONS

1. Robert E. Bartholomew et al., "Mass Psychogenic Illness and the Social Network: Is It Changing the Pattern of Outbreaks?," *Journal of the Royal Society of Medicine* 105 (2012): 509–12; Donna M. Goldstein and Kira Hall, "Mass Hysteria in Le Roy, New York," *American Ethologist* 42 (2015): 640–57; Susan Dominus, "What Happened to the Girls in Le Roy," *New York Times*, March 7, 2012.

2. L. L. Langness, "Hysterical Psychosis: The Cross-Cultural Evidence," *American Journal of Psychiatry* 124 (Aug. 1967): 143–52.

3. Adam Smith, *The Theory of Moral Sentiments* (1759; New York: Augustus M. Kelley, 1966).

4. Frederique de Vignemont and Tania Singer, "The Empathic Brain: How, When, and Why?," *Trends in Cognitive Sciences* 10 (2006): 435–41.

5. Elaine Hatfield et al., "Primitive Emotional Contagion," *Review of Personality and Social Psychology* 14 (1992): 151–77.

6. W. S. Condon and W. D. Ogston, "Sound Film Analysis of Normal and Pathological Behavior Patterns," *Journal of Nervous Mental Disorders* 143 (1966): 338–47.

7. James H. Fowler and Nicholas A. Christakis, "Dynamic Spread of Happiness in a Large Social Network: Longitudinal Analysis over 20 Years in the Framingham Heart Study," *BMJ* 337 (2008): a2338.

8. Adam D. I. Kramer, Jamie E. Guillory, and Jeffrey T. Hancock, "Experimental Evidence of Massive-Scale Emotional Contagion Through Social Networks," *Proceedings of the National Academy of Sciences* 111 (2014): 8788–90.

9. Emilio Ferrara and Zeyao Yang, "Measuring Emotional Contagion in Social Media," *PLoS One* 10 (2015): e0142390.

10. Allison A. Appleton and Laura D. Kubzansky, "Emotion Regulation and Cardiovascular Disease Risk," in *Handbook of Emotion Regulation*, ed. J. J. Gross (New York: Guilford Press, 2014), 596–612.

11. James Stockdale, "Tranquility, Fearlessness, and Freedom" (a lecture given to the Marine Amphibious Warfare School, Quantico, Va., April 18, 1995); "Vice Admiral James Stockdale," obituary, *Guardian*, July 7, 2005.

12. Epictetus, *The Enchiridion* (New York: Dover, 2004), 6.

13. Ibid., 1; note that "control" is translated here as "power."

14. J. McMullen et al., "Acceptance Versus Distraction: Brief Instructions, Metaphors, and Exercises in Increasing Tolerance for Self-Delivered Electric Shocks," *Behavior Research and Therapy* 46 (2008): 122–29.

15. Amit Etkin et al., "The Neural Bases of Emotion Regulation," *Nature Reviews Neuroscience* 16 (2015): 693–700.

16. Grace E. Giles et al., "Cognitive Reappraisal Reduces Perceived

Exertion During Endurance Exercise," *Motivation and Emotion* 42 (2018): 482–96.

17. Mark Fenton-O'Creevy et al., "Thinking, Feeling, and Deciding: The Influence of Emotions on the Decision Making and Performance of Traders," *Journal of Organizational Behavior* 32 (2010): 1044–61.

18. Daniel Kahneman, *Thinking, Fast and Slow* (New York: Farrar, Straus and Giroux, 2011).

19. Matthew D. Lieberman et al., "Subjective Responses to Emotional Stimuli During Labeling, Reappraisal, and Distraction," *Emotion* 11 (2011): 468–80.

20. Andrew Reiner, "Teaching Men to Be Emotionally Honest," *New York Times*, April 4, 2016.

21. Matthew D. Lieberman et al., "Putting Feelings into Words," *Psychological Science* 18 (2007): 421–28.

22. Rui Fan et al., "The Minute-Scale Dynamics of Online Emotions Reveal the Effects of Affect Labeling," *Nature Human Behaviour* 3 (2019): 92.

23. William Shakespeare, *Macbeth*, act 4, scene 3.

Index

Leonard Mlodinow received his PhD in theoretical physics from the University of California, Berkeley, was an Alexander von Humboldt fellow at the Max Planck Institute, and was on the faculty of the California Institute of Technology. His previous books include the best sellers *The Drunkard's Walk* (a *New York Times* Notable Book of the Year), *The Grand Design* and *A Briefer History of Time* (both with Stephen Hawking), *Subliminal* (winner of the PEN/E. O. Wilson Literary Science Writing Award), and *War of the Worldviews* (with Deepak Chopra), as well as *Stephen Hawking: A Memoir of Friendship and Physics*, *Elastic*, *The Upright Thinkers*, *Feynman's Rainbow*, and *Euclid's Window*.

A NOTE ON THE TYPE

This book was set in Janson, a typeface long thought to have been made by the Dutchman Anton Janson, who was a practicing typefounder in Leipzig during the years 1668–1687. However, it has been conclusively demonstrated that these types are actually the work of Nicholas Kis (1650–1702), a Hungarian, who most probably learned his trade from the master Dutch typefounder Dirk Voskens. The type is an excellent example of the influential and sturdy Dutch types that prevailed in England up to the time William Caslon (1692–1766) developed his own incomparable designs from them.

Typeset by Scribe,
Philadelphia, Pennsylvania

Printed and bound by Berryville Graphics,
Fairfield, Pennsylvania

Designed by Soonyoung Kwon